NMR and its Applications to Living Systems

SECOND EDITION

DAVID G. GADIAN

Institute of Child Health, London

OXFORD UNIVERSITY PRESS

Oxford New York Tokyo

1995

Oxford University Press, Walton Street, Oxford OX2 6DP
Oxford New York
Athens Auckland Bangkok Bombay
Calcutta Cape Town Dar es Salaam Delhi
Florence Hong Kong Istanbul Karachi
Kuala Lumpur Madras Madrid Melbourne
Mexico City Nairobi Paris Singapore
Taipei Tokyo Toronto
and associated companies in
Berlin Ibadan

Oxford is a trade mark of Oxford University Press

Published in the United States
by Oxford University Press Inc., New York

A catalogue record for this book is available from the British Library

Library of Congress Cataloging in Publication Data
Data available

ISBN 0 19 855281 5 (Hbk)
ISBN 0 19 855803 1 (Pbk)

Typeset by Colset Pte Ltd, Singapore.

Printed and bound in Great Britain by
Biddles Ltd, Guildford and King's Lynn

To Ro

Preface

Nuclear magnetic resonance (NMR) is now extensively used as a non-invasive means of obtaining clinical images and of studying tissue metabolism *in vivo*. At the time of writing, over 6000 whole-body NMR systems are available for clinical imaging; some of these systems are being used for studies of human metabolism, while smaller scale instruments are suitable for more basic research studies. There have been many remarkable developments in this field of research in recent years, and I therefore felt that it would be a good time to rewrite the first edition of my book, many aspects of which are now, to say the least, out of date. In view of the nature of these developments, I felt it appropriate to focus on applications in humans, but I also emphasize how basic NMR research studies can complement and aid interpretation of clinical findings.

Over the years, it has become convenient to divide *in vivo* NMR into two branches: magnetic resonance imaging (MRI, based primarily on the detection of signals from water), and magnetic resonance spectroscopy (MRS, as used for the detection of metabolites). However, the techniques of MRI and MRS have many features in common with each other, and they are really variants or extensions of the more traditional use of NMR as a branch of spectroscopy. In view of this, and in view of the ways in which MRI and MRS have been evolving, I believe that it is now timely to emphasize the interrelationships between imaging and spectroscopy, rather than their differences, and this is the approach that I have tried to adopt in this book. I have also tried as far as possible to write the book at a level that is appropriate for newcomers to the field, and for users of NMR who may not have a strong background in the physical sciences.

Where appropriate, I have followed the format of the first edition. Chapter 1 gives an introduction to NMR and its applications to living systems, and includes a brief historical account of the development of this area of research. This is followed by three chapters on applications of the techniques. These chapters outline the type of information that MRI and MRS can provide, and give illustrative examples that serve to emphasize the scope (and limitations) of the techniques that are currently available. The MRI applications focus on physiology (including the new techniques of functional imaging) rather than on diagnostic radiology, which is a vast area that is well covered by numerous other texts. The latter half of the book is concerned with theoretical and technical aspects of NMR as applied to living systems, with chapters on the basic principles of NMR, the parameters that characterize NMR signals, instrumentation, and pulse sequences.

I would like to take this opportunity to acknowledge my debt to all the many colleagues and friends with whom I have worked over the last twenty years, initially in Oxford and subsequently in London. I hope they will forgive me if I don't mention them by name; I owe so much to them, and a long list would do little justice to them all. I also greatly appreciate the many world-wide friendships I have developed through working in this exciting area of research.

For the writing of this second edition, I would like to thank Alan Connelly, Alistair Howseman, Steve Williams, Nick Preece, and Albert Busza for their help with various aspects of the manuscript. Special thanks are due to Sally Dowsett for all her work with the text, and in particular with the diagrams; without her, no doubt I would have missed even more deadlines. I would also like to acknowledge everyone who has granted permission to reproduce diagrams, and all those who have submitted figures to me. I received many valuable comments on the first edition, and I have rewritten the book taking these comments into account. I hope that in so doing I have preserved the useful parts of the first edition, and that at the same time I have managed to do some justice to the remarkable developments that have taken place since the first edition initially appeared.

London   D. G. G.
January 1995

Contents

The plates referred to on pages 71, 127, and 130 follow page 84.

1
Introduction

1.1 A NON-INVASIVE METHOD OF STUDYING LIVING SYSTEMS

There are clear attractions in using non-invasive methods for the study of living systems. The use of such methods to image the human body has obvious importance in diagnostic medicine. More generally, there are many areas of biomedical research where non-invasive studies can add significantly to the information provided by more traditional, invasive techniques. For example, if we wish to study the metabolic basis of disease, then the analysis of body fluids, biopsy samples, and model systems (e.g. cultured cells) can only provide an incomplete picture of the abnormalities occurring within diseased tissue; ultimately it is important to measure directly and non-invasively the metabolic state of the tissue itself.

Similarly, our understanding of metabolic pathways in normal tissue, and of the mechanisms whereby cellular metabolism is regulated, has relied to a large extent on measurements made *in vitro*. *In vitro* studies have the virtues of simplicity and of enabling the experimenter to control at will the conditions under which the pathway is studied. However, any conclusions that are reached must eventually be confirmed by observations made *in vivo*.

Nuclear magnetic resonance (NMR) is now widely used as a non-invasive means of obtaining clinical images and of studying tissue metabolism *in vivo*. In this book, we discuss the principles of NMR, the ways in which NMR can be used to study living systems, and the scope and limitations of the method. The applications of NMR are extensive and diverse. For example, at the time of writing, over 6000 whole-body systems are available for clinical imaging; some of these systems are being used for studies of human metabolism, while smaller scale instruments are suitable for more basic research on biopsy specimens, body fluids, cultured cells, isolated tissues, and small animals. While we focus in this book on applications in humans, we also emphasize how basic NMR research studies can complement and aid interpretation of clinical findings.

For some years, the imaging applications of NMR evolved more or less independently of the metabolic applications, and it became convenient to introduce the two designations of MRI (magnetic resonance imaging, based primarily on the detection of signals from water) and MRS (magnetic resonance spectroscopy, as used for the detection of metabolites). However, the techniques of MRI and MRS have many features in common with each other, and they are really variants, or extensions, of the more traditional use of NMR as a branch of spectroscopy. Moreover, they are inextricably linked by the new generation of NMR techniques that combine spatial and chemical information. In addition, it is apparent that MRI is not simply a method of obtaining diagnostic radiological images; it can also provide a great deal of physiological information. For example, functional neuroimaging, based on various properties of the NMR signal from water, appears to have a very bright future. Therefore, although it is often convenient to use the labels of MRI and MRS, it is important to appreciate that there is very considerable overlap between these two magnetic resonance approaches. As far as possible, therefore, we emphasize in this book the strong interrelationships that exist between imaging and spectroscopy.

The main purpose of this introductory chapter is to provide a brief overview of the material that is covered in subsequent chapters. We begin by giving an introduction to the field of spectroscopy in general, and to NMR in particular. We then briefly discuss the type of information that NMR can provide about biological systems, and conclude the chapter by outlining the historical development of NMR as a method of studying living systems.

1.2 AN INTRODUCTION TO SPECTROSCOPY

In 1666 Isaac Newton obtained a 'triangular glass prism to try therewith the celebrated phenomena of colours'. He showed that the prism separated white light into what he called a 'spectrum' of the colours, and his experiments marked the beginnings of the branch of science that we term spectroscopy. Spectroscopy deals with the interaction of electromagnetic radiation with matter. The radiation that is absorbed or emitted by the sample under investigation is detected in some way, and the output of the detector, displayed as a function of the frequency or wavelength of the radiation, is called a spectrum. The spectrum can be interpreted in terms of some aspect of the atomic or molecular structure of the sample.

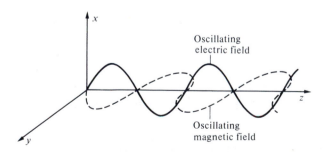

Fig. 1.1 Electromagnetic radiation travelling in the z-direction. The radiation illustrated here is plane polarized as the electric field is always in the xz-plane.

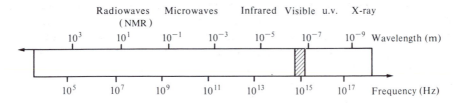

Fig. 1.2 Regions of the electromagnetic spectrum. The visible region occupies a small band around 10^{15} Hz, as shown by the shaded area.

The electromagnetic theory developed by James Clerk Maxwell over a hundred years ago provided an elegant synthesis of a wide range of phenomena involving electric and magnetic fields. In particular, his theory provided an explanation of the wave propagation of light, for it predicted the existence of electromagnetic disturbances that should travel with a speed equal to the accepted value of the speed of light. These disturbances consist of coupled oscillating electric and magnetic fields that are perpendicular to each other and to the direction of propagation of the radiation (Fig. 1.1). The frequency of oscillation and the wavelength of the radiation are related by the equation

$$c = \nu\lambda \tag{1.1}$$

where c is the speed of the radiation (commonly known as the speed of light), ν is the frequency, and λ is the wavelength. The positions of spectral lines can be characterized by their frequency, wavelength, or wave number $1/\lambda$, all of which are equivalent as they are related to each other by eqn (1.1). The spectral regions of electromagnetic radiation are shown in Fig. 1.2, from which it can be seen that visible light occupies just a small fraction of the overall range of frequencies.

Electromagnetic radiation can also be regarded as consisting of discrete packets, or quanta, of energy that travel with the speed of light. The fundamental relationship that links the quantal and wave-like nature of radiation is

$$\varepsilon = h\nu \tag{1.2}$$

where ε is the energy of the quanta, ν is the frequency of the associated radiation, and h is the Planck constant. The interaction of a molecule with radiation involves the absorption or emission by the molecule of a quantum of radiation, which is accompanied by a transition from one energy level to another. Since energy is conserved, the energy difference, ΔE, between the two levels must be equal to the energy of the quantum of radiation, i.e.

$$\Delta E = h\nu \tag{1.3}$$

where ν is the frequency of the radiation that is absorbed or emitted. The various forms of spectroscopy can therefore be classified according to the frequency of the radiation or by the type of transition that is involved. For example, the absorption or emission of radiation in the X-ray region is accompanied by changes in the energy of the inner electrons of atoms or molecules. The visible and ultraviolet regions are associated with transitions of the valence electrons. At lower frequencies, the microwave and infrared regions are characteristic of molecular rotational and vibrational energy changes, respectively. The frequencies characteristic of NMR are low, typically in the range 1–750 MHz, and are therefore (from eqn (1.3)) associated with transitions between energy levels that are relatively closely spaced. These levels correspond to different magnetic states of atomic nuclei, as discussed in the next section.

In considering the use of these various forms of spectroscopy for the investigation of biological material, it is important to make a general point about the interaction of radiation with tissue. If this interaction is too weak, then the tissue will be completely transparent to the radiation, and it will not be possible to derive useful information about the tissue. However, if the interaction is too strong, it may produce unacceptable dangers, and, furthermore, the radiation may be fully absorbed by superficial regions, making it impossible to investigate internal regions of the body. It turns out that, on these criteria alone, only certain frequency 'windows' are appropriate for use in humans, one of these being the radiofrequencies characteristic of NMR.

1.3 THE BASIS OF NMR

Certain atomic nuclei, such as the hydrogen nucleus, ^1H, or the phosphorus nucleus, ^{31}P, possess a property known as spin. This can be visualized as a spin-

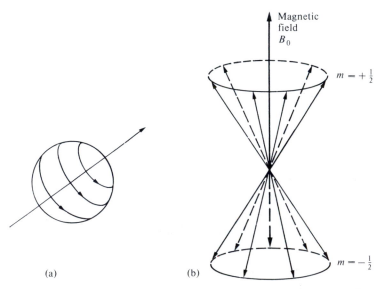

(a)   (b)

Fig. 1.3 (a) A nucleus can be visualized as spinning about its own axis, which is the axis of its magnetic moment (see Section 5.1). (b) Orientations that can be taken up in an applied field B_0 by the magnetic moment of a nucleus of spin $\frac{1}{2}$. The orientations are specified by the quantum number m (see Section 5.2.1) and describe two cones.

ning motion of the nucleus about its own axis (Fig. 1.3(a)). Associated with the spin is a magnetic property, so that the nucleus can be regarded as a tiny bar magnet with its axis along the axis of rotation. If a static magnetic field B_0 is applied to a sample containing such nuclei (e.g. to water in the case of ^{1}H NMR), we might expect the nuclear magnets to align along the field, just as a compass needle aligns along a magnetic field. However, the nuclei have spin and obey the laws of quantum mechanics, and therefore they do not behave like conventional bar magnets. Instead, we find that nuclei such as ^{1}H that have a spin quantum number $I = \frac{1}{2}$ (see Section 5.2.1) can have one of two orientations with respect to the applied field, as shown in Fig. 1.3(b). These two orientations have slightly different energies, and the energy difference between the two states is proportional to the magnitude of the applied field (see Fig. 1.4). Transitions between these states can be induced by applying an additional oscillating magnetic field.[1] This oscillating field, B_1, which is applied in the plane perpendicular to the direction of B_0, has a frequency ν_0 that satisfies eqn (1.3), i.e.

$$\Delta E = h\nu_0 \qquad (1.4)$$

where ΔE is the energy separation of the levels. We then find (see Section 5.2.3) that

$$\nu_0 = \frac{\gamma}{2\pi} B_0 \qquad (1.5)$$

where B_0 is the magnitude of the static magnetic field, γ is known as the gyromagnetic (or magnetogyric) ratio of the nucleus, and ν_0 is the resonance

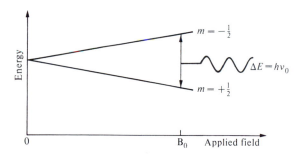

Fig. 1.4 Energies of the two orientations of a nucleus of spin $\frac{1}{2}$ plotted as a function of the applied field B_0.

[1] Thus, strictly speaking, NMR does not (as is commonly believed) utilize electromagnetic radiation which, as we have seen, involves coupled electric and magnetic fields; only the magnetic component is present in NMR studies.

frequency. The ratios γ varies from one nuclear isotope to another, and this is why ^1H, ^{13}C, and ^{31}P NMR, for example, are all performed at different frequencies in a given field. (^1H NMR is commonly known as proton NMR because the hydrogen nucleus consists of a single proton.)

The nuclear magnets interact very weakly with the field B_0, and this accounts for the low values of the energy separation ΔE and of the characteristic NMR frequencies. We shall see in Section 5.2.4 that this weakness is also responsible for the low sensitivity of NMR.

1.4 THE NMR NUCLEI

Only those nuclei that have magnetic properties, i.e. those that possess non-zero spin (see Chapter 5), give rise to NMR signals. The most biologically relevant of these nuclei, together with some of their properties, are listed in Table 1.1. Nuclei of spin $\frac{1}{2}$ tend to have better NMR characteristics than those of spin greater $\frac{1}{2}$, and this is why most studies of living systems use nuclei of spin $\frac{1}{2}$, such as ^1H, ^{13}C, ^{19}F, and ^{31}P. The abundant isotopes of carbon and oxygen, ^{12}C and ^{16}O, have zero spin and therefore do not produce NMR signals.

1.5 NMR SPECTROSCOPY AND IMAGING

The resonance frequency of a nucleus is directly proportional to the *local* magnetic field experienced by the nucleus. In Section 1.3, we considered this

Table 1.1 NMR properties of the nuclei commonly used in biology

Nucleus	Spin quantum number	Resonance frequency at 2.35T, in MHz	Natural abundance (%)	Relative sensitivity at constant field [1]
^1H	1/2	100.0	99.98	100
^2D	1	15.4	0.0156	1.5×10^{-4}
^{13}C	1/2	25.1	1.1	1.8×10^{-2}
^{14}N	1	7.2	99.6	1.0×10^{-1}
^{15}N	1/2	10.1	0.37	3.8×10^{-4}
^{19}F	1/2	94.1	100.0	83.0
^{23}Na	3/2	26.5	100.0	9.3
^{31}P	1/2	40.5	100.0	6.6
^{35}Cl	3/2	9.8	75.4	3.5×10^{-1}
^{39}K	3/2	4.7	93.1	4.7×10^{-2}

[1] Relative sensitivity is the NMR sensitivity of the nucleus relative to that of an equal number of protons, multiplied by the percentage natural abundance. As discussed in Section 1.8, numerous additional factors also influence the signal-to-noise ratios obtained in spectra or images.

field to be the applied field B_0. If this were true, and if the field were perfectly uniform over the entire volume of the sample, then all protons, for example, would absorb energy at the same frequency. Under such circumstances, NMR would be a relatively uninteresting and uninformative technique, incapable of attributing protons to different chemical or spatial environments. In practice, however, there are frequency shifts associated with field variations; these are exploited in spectroscopy to provide chemical information and in imaging to provide spatial information. We first discuss the basic characteristics of NMR spectra, before showing how analogous concepts apply to imaging.

1.5.1 Spectroscopy

The application of a static field B_0 induces electronic currents in atoms and molecules, and these produce a further small field at the nucleus which is proportional to B_0. The total effective field, B_{eff}, at the nucleus can therefore be written

$$B_{\text{eff}} = B_0 (1 - \sigma) \tag{1.6}$$

where σ expresses the contribution of the small secondary field generated by the electrons. Using eqn (1.5) we find that

$$\nu_0 = \frac{\gamma}{2\pi} B_0 (1 - \sigma) \tag{1.7}$$

σ is a dimensionless constant, known as the shielding or screening constant, and has values typically in the region 10^{-6}–10^{-3}. The magnitude of σ is dependent upon the electronic environment of the nucleus, and therefore nuclei in different chemical environments give rise to signals at different frequencies. The separation of resonance frequencies from an arbitrarily chosen reference frequency is termed the *chemical shift*, and is expressed in terms of the dimensionless units of parts per million (ppm). A simple illustration of the chemical shift is given by the ^{1}H NMR spectrum of acetic acid shown in Fig. 1.5. The CH_3 and COOH protons experience different chemical environments and therefore give rise to two separate signals or resonances. The *intensities* of the two signals, as measured from their *areas*, are proportional to the number of nuclei that contribute towards them, and so the relative areas of the signals in Fig. 1.5 are 3:1. However, it should be emphasized that other factors, including the effects of *spin–lattice relaxation* and *spin–spin relaxation*, can also influence the signal intensities. As discussed in the following sections, such effects are particularly important for the generation of contrast in MRI. In spectroscopy, however, while relaxation effects can sometimes be used for selective detection or suppression of specific signals, they may equally be regarded as a nuisance that complicates the measurement of metabolite concentrations.

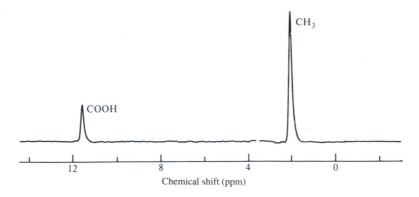

Fig. 1.5 ^{1}H NMR spectrum of acetic acid ($CH_3 COOH$). The relative areas of the two signals are 3:1. The frequencies of the signals are expressed in terms of ppm relative to the signal from the reference compound tetramethylsilane (TMS).

The ^{31}P spectrum of adenosine triphosphate (ATP), shown in Fig. 1.6, contains three groups of spectral lines corresponding to the α-, β-, and γ-phosphates of ATP, all of which are chemically different. In addition, each group is split into a multiplet of lines as a result of an interaction between the neighbouring phosphorus nuclei, which is known as *spin–spin coupling*. This interaction is transmitted through electronic bonds, and the magnitude of the splitting is independent of the applied field B_0.

ATP in tissues also gives rise to these three groups of ^{31}P signals, although the spectral resolution *in vivo* is often insufficient to resolve the splittings within each group. Such signals, together with additional ^{31}P signals from other metabolites, form the basis of many NMR studies of metabolism *in vivo*. For example, changes in the concentrations of ATP and/or other metabolites can be monitored non-invasively through changes in the intensities of their respective signals.

The various factors that influence the characteristics of NMR spectra are discussed in much greater detail in Chapter 6.

1.5.2 Imaging

Although the practical details of magnetic resonance imaging may be fairly complex, the basic principles are quite straightforward. If we consider a rectangular vessel of water in a uniform field B_0 (Fig. 1.7(a)), then all the water protons will give rise to a signal of the same frequency (Fig. 1.7(b)), given by eqn (1.5). Let us now superimpose a small linear field gradient upon B_0, so that the field is greater at the left side of the vessel than at the right (Fig. 1.7(c)). Since the resonance frequency of the protons is directly proportional to the field that they experience, the water at the left will generate a signal of higher frequency than the water at the right. An intermediate

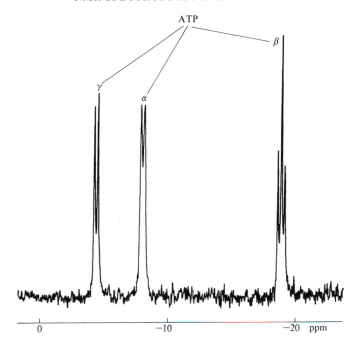

Fig. 1.6 ^{31}P NMR spectrum obtained at 74 MHz from a solution of ATP at pH 7.0. The chemical shifts in this spectrum, and in most of other ^{31}P spectra in this book, are expressed relative to the signal of phosphocreatine at pH 7, which is assigned the value 0 ppm (see Section 6.1.2 for a discussion of frequency standards).

frequency will be detected from water in the centre. The resulting signal will therefore consist of a superposition of a continuous range of slightly shifted resonances, the overall effect being that a rectangular-shaped signal is observed (Fig. 1.7(d)). If a plastic strip containing no detectable protons were placed in the water (Fig. 1.7(e)), the resulting signal would be affected as shown in Fig. 1.7(f). Clearly, these NMR signals provide information about the spatial distribution of water within the sample.[1] It should be emphasized that the display of the NMR signals in Fig. 1.7 is exactly analogous to that of the spectra in Figs. 1.5 and 1.6, except that the frequency now encodes for space rather than chemistry.

It is apparent, however, that a one-dimensional profile or projection of the type shown in Fig. 1.7(f) provides a far from complete description of the structure of a three-dimensional sample; similar signals could be obtained from samples of a variety of shapes. The required extension to two (or three) dimensions involves the collection of NMR data that are obtained with a range

---

[1] For simplicity, the word 'sample' is commonly used in this book to represent the system under investigation, whether this be a suspension of cells, a tissue extract, or a human subject.

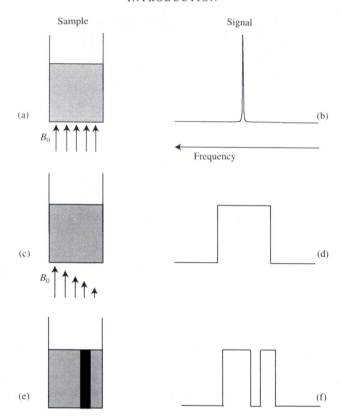

Fig. 1.7 Diagram illustrating one-dimensional imaging. In a uniform magnetic field, all of the water protons in a rectangular vessel give rise to a signal of the same frequency ((a) and (b)). If NMR signals are acquired in the presence of a linear field gradient so that the field is greater at the left side of the vessel than at the right (c), then the resulting spectrum will be shaped like a rectangle (d). If a plastic strip containing no NMR-detectable protons is placed in the vessel (e), then the signal will be modified as shown in (f). The spectra in (d) and (f) represent one-dimensional projections of proton density along the direction of the gradient, and are effectively one-dimensional images. Note that the magnitude of the field gradient shown schematically under (c) is greatly exaggerated; typical values for the static field B_0 and for the gradient strengths are 1.5 T and 5mT m^{-1} respectively; i.e. the introduction of the field gradient changes the net field by only a small percentage.

of different field gradient conditions (see Sections 5.7.2 and 8.8). The question arises as to how to display the eventual output. Clearly, it makes sense to display the signal in such a way that it gives a visual representation of the object, and so a two-dimensional cross-section is most simply displayed as a grid of data points or picture elements (pixels), in which the brightness of each element represents the signal intensity (Fig. 1.8). Such cross-sectional images are generated by computer analysis of the data, analogous with image reconstruction

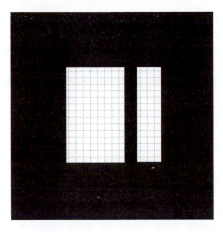

Fig. 1.8 A two-dimensional cross-sectional image of the specimen shown in Fig. 1.7(e). The signals are represented as a grid of data points or picture elements, the brightness of each element representing the signal intensity.

in X-ray scanning. In comparing magnetic resonance imaging with X-ray scanning, one of the features of MRI is that the gradients can be applied equally easily along any of the axes, so that images in each of the corresponding planes can be obtained directly. Another important point is that, as in spectroscopy, the signal intensities in magnetic resonance images are influenced not only by the concentrations of nuclei that contribute towards them, but also by several other factors, including the effects of relaxation. These effects can be exploited in order to optimize image contrast.

The display of NMR data becomes more complex as the information content increases. For example, in spectroscopic imaging, a large number of spectra are obtained simultaneously from many different regions; such spectra are commonly displayed in the form of separate images for each metabolite, although detailed analysis will still require separate viewing of each of the individual spectra. Similarly, it may be desirable to overlay spectral, functional, or flow data on conventional images, in which case colour displays may be desirable. Three-dimensional data acquisition requires further computer strategies for optimal display of structures in all three dimensions.

1.5.3 Localized spectroscopy

We have seen that in spectroscopy the frequencies of NMR signals provide chemical information, whereas in imaging they provide spatial information. Commonly, it is necessary to obtain spectra from well-defined localized regions of the body, in which case a combination of imaging and spectroscopy techniques may be required.

The initial description of localized spectroscopy for studies of tissue metabolism did not, in fact, utilize imaging techniques. Instead, use was made of the localizing properties of a particular type of radiofrequency transmit/receive coil termed a 'surface coil'. In its simplest form, a surface coil consists of a loop of wire placed adjacent to the selected region of interest. Such a coil detects signals primarily from the region immediately in front of it. While the localization is far from perfect, it does provide a simple and effective means of investigating superficial tissue, and is particularly useful for studies of skeletal muscle metabolism.

It soon became apparent, however, that more sophisticated methods of spectral localization were needed and, as discussed in Sections 8.9 and 8.10, two general approaches have evolved. One involves the detection of signals from a single well-defined region of the body, while the other involves the simultaneous acquisition of data from many different regions. Both approaches use magnetic field gradients, as in imaging, to obtain spatial information, but are also able to retain the required chemical information.

1.6 RELAXATION

It was mentioned above that the effects of relaxation can influence the appearance of spectra and of images. In this section we briefly introduce the phenomenon of relaxation; a much more detailed description is given in Section 6.3.

When a sample containing nuclei of spin $\frac{1}{2}$ is placed in a magnetic field B_0, the nuclear magnets take up one of two orientations with respect to the field (see Section 1.3), with a slight excess in the orientation of lower energy. As a result, the sample becomes slightly magnetized, the net component of magnetization being along the direction of the field (Fig. 1.9(a)). The radiofrequency field B_1 that is used for signal detection causes the magnetization to be perturbed; for example a 90° radiofrequency pulse causes the magnetization to be tilted into the plane perpendicular to B_0 (Fig. 1.9(b)). Following this perturbation, processes take place whereby the nuclear magnetization returns to its initial equilibrium value. These processes are characterized by two relaxation times, the spin–lattice relaxation time T_1 and the spin–spin relaxation time T_2. T_1 is the time constant for the recovery of magnetization along the direction of B_0, while T_2 is the time constant for the decay of magnetization in the plane perpendicular to B_0.

As a result of the variations in water T_1 and T_2 that exist, firstly between different tissues and secondly between normal and diseased tissue, much of the contrast in MRI is generated by T_1- and T_2-dependent effects. The degree of contrast is strongly influenced by the manner in which the radiofrequency field is applied, in particular by the nature of the radiofrequency pulse sequences that are used. For example, different sequences can give predomi-

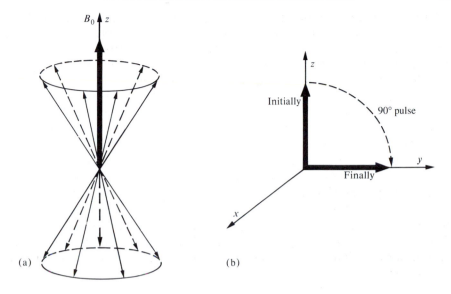

Fig. 1.9 (a) Prior to the application of a radiofrequency pulse, the net component of magnetization, indicated by the heavy arrow, is along the direction of B_0 (i.e. the z-axis). (b) The effect of a 90° pulse is to tilt the magnetization into the xy-plane.

nantly T_1-weighting, predominantly T_2-weighting, or a complex mixture of the two. Such pulse sequences are discussed briefly in Section 1.10 and in more detail in Chapter 8.

1.7 LINEWIDTHS AND SPECTRAL RESOLUTION

Signals from molecules in solution often have a characteristic lineshape $g(v)$ given by

$$g(v) \propto \frac{T_2}{1 + 4\pi^2 T_2^2 (v - v_0)^2} \tag{1.8}$$

where v_0 is the resonance frequency, and T_2 is the spin–spin relaxation time. This is known as a Lorentzian lineshape, and is illustrated in Fig. 1.10. The intensity of the signal is given by the shaded area. The natural *linewidth* $\Delta v_{1/2}$ at half-height is given by

$$1/T_2 = \pi \Delta v_{1/2} \tag{1.9}$$

In general, as molecules become increasingly immobilized they produce broader signals. Therefore, spectra of living systems reveal narrow signals from metabolites which have a high degree of molecular mobility, whereas macromolecules, which are highly immobilized (such as DNA and membrane

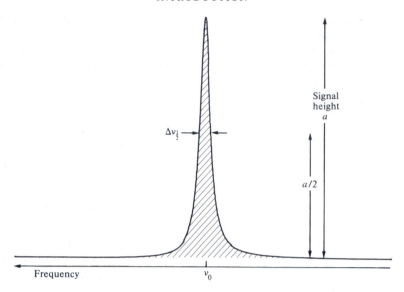

Fig. 1.10 The Lorentzian lineshape. $\Delta v_{1/2}$ is the natural linewidth determined by T_2; the observed linewidth is often greater because of effects such as B_0 inhomogeneity. The intensity of the signal is given by the shaded area, and the signal height, or amplitude, is equal to a.

phospholipids), produce very much broader signals which are either invisible or appear as broad components underlying the signals from the metabolites. However, linewidths and lineshapes can be influenced not just by molecular mobility, but also by a range of other factors including *chemical exchange* (see Section 6.4) and magnetic field inhomogeneities.

Magnetic field homogeneity is a critical feature of NMR technology, particularly in spectroscopy studies. The reason for this is apparent from eqn (1.5), from which it can be seen that any inhomogeneity ΔB_0 will broaden resonances by an amount $v_0 = (\gamma/2\pi)\Delta B_0$; in other words, a magnetic field inhomogeneity of, say, 10 ppm over the volume of interest will broaden each signal by 10 ppm. Since 10 ppm represents more or less the whole spectral range in ^{1}H NMR spectroscopy, under these circumstances none of the resonances would be resolvable from any of the others. Clearly, therefore, the field homogeneity that is required is dictated by the degree of spectral resolution needed. ^{1}H NMR spectroscopy imposes particularly stringent requirements; for ^{1}H studies of brain metabolism, it is desirable for the field to be homogeneous to within 3 parts in 10^8. High-field spectrometers that are used for studies of solutions may have field homogeneity as remarkable as 1 part in 10^9, although of course this is over a much smaller sample volume (e.g. 0.5 ml) than the volumes characteristic of *in vivo* studies.

Because of this need for high field homogeneity, the acquisition of spectroscopy data (and in some cases of imaging data also) is preceded by the procedure known as 'shimming', which involves adjustments that are designed to optimize the field homogeneity for each given study. However, regardless of the success of this procedure, field inhomogeneities will inevitably make some contribution to the observed linewidth. The observed linewidth Δv_{obs} will therefore be greater than the natural linewidth of eqn (1.9), and by analogy with eqn (1.9) we can write

$$1/T_2^* = \pi \Delta v_{obs} \qquad (1.10)$$

where T_2^* differs from T_2 in that it incorporates the effects of field inhomogeneities as well as intrinsic relaxation effects.

1.8 SENSITIVITY AND SPATIAL RESOLUTION

A major disadvantage of NMR is its inherent lack of sensitivity, which can be expressed in terms of the signal-to-noise ratio of the spectral lines or pixels (see Fig. 1.11). The signal-to-noise ratio is dependent upon a wide range of factors including:

(1) the nucleus under investigation (see Table 1.1 for relative sensitivities);
(2) the volume (voxel size) of the region that contributes to each signal, and the fractional volume of the region of interest compared to that 'seen' by the radiofrequency coil(s);
(3) the magnetic field strength B_0;
(4) the design and performance of the instrument, including the radiofrequency coil(s);
(5) the time for which data are accumulated;
(6) the widths of the signals (which in imaging are related to the gradient strengths);
(7) the relaxation times T_1 and T_2;
(8) the radiofrequency pulse sequence that is used for data collection;
(9) the extent of isotopic enrichment, where this is feasible; and of course
(10) the concentration of the nuclei under investigation.

The poor sensitivity of NMR imposes limitations on the concentrations of compounds that can be detected, and upon the spatial resolution that can be achieved. Because of the large number of variables given above, it is difficult to give anything other than an order-of-magnitude estimate for the concentrations that are required and for the spatial resolution that can be achieved. Typically, however, we can anticipate that, for metabolic studies *in vivo*, minimum concentrations of 0.2 mM and above will be required in order for a metabolite to give a detectable signal. Thus, most spectroscopy studies are concerned with the detection of metabolites such as ATP, phosphocreatine, and inorganic phosphate that have concentrations in the range of 1 mM and above. For studies of the human brain, localized ^1H spectroscopy (or chemical shift imaging) of metabolites in this concentration range can be carried out with

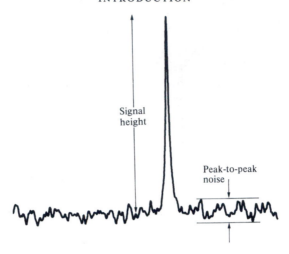

Signal
height

Peak-to-peak
noise

Fig. 1.11 Signal and noise in an NMR spectrum. The signal-to-noise ratio can be defined in a number of ways; a simple definition is signal amplitude (or in images the pixel intensity) divided by the standard deviation of the background noise.

linear resolution of about 1–2 cm; it is difficult to achieve better resolution than this simply because smaller volume elements do not give rise to signals that are large enough to be detected above the background noise. The linear resolution in conventional MRI is an order-of-magnitude or more superior to this because the water protons are present at far higher concentrations; the concentration of water in tissues is about 40 M and therefore sufficient signal intensity can be observed from very much smaller volume elements.

Fortunately, the spatial resolution available for animal and solution studies scales approximately with the size of the animal or specimen, because of the effects of other variables in the list given above. The concentration requirements for small animal studies are much the same as for humans, while for solutions there is scope in high-field systems for detecting concentrations considerably below 0.2 mM, particularly for ^{1}H and ^{19}F studies.

1.9 DETECTION OF NMR SIGNALS

Figure 1.12 shows a very simple block diagram of the instrumentation required for the detection of NMR signals. The transmitter drives a radiofrequency current through a transmitter coil, and thereby generates oscillating magnetic fields of the appropriate timing, power, and frequency within the sample. The nuclei absorb radiofrequency energy at their resonance frequencies, and the ensuing signal is fed by a receiver coil (which in practice may be a combined transmit/receive coil) into the receiver, where it is analysed and processed in the required way, using a dedicated computer. Chapter 7 deals in some detail

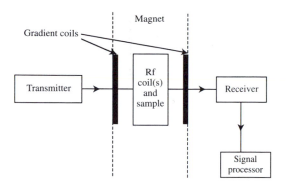

Fig. 1.12 A simple block diagram of an NMR instrument.

with the various units that make up an NMR instrument and with the procedures that are commonly used for the acquisition of images or spectra.

Since NMR relies on the detection of signals covering a range of frequencies, one approach to the collection of signals might be to sweep the applied irradiation through the required range of frequencies. Equivalently, the frequency could be kept constant, and a magnetic field sweep could be employed. This approach formed the basis of the continuous-wave mode of detection, which was used in the early days of NMR spectroscopy. However, continuous-wave NMR was superceded by the development in the 1960s of Fourier-transform NMR, and it is this latter mode of detection that forms the basis of modern NMR technology and techniques.

In Fourier-transform NMR, the radiofrequency field that is used for excitation of signals is applied in the form of pulses, the duration of which may vary from a few microseconds to many milliseconds. The bandwidth of these pulses (i.e. their effective spread in frequency) is sufficiently large to excite all of the nuclei within the required frequency range. In comparison with continuous-wave NMR, this produces a considerable improvement in sensitivity, because signals are detected simultaneously rather than one by one. The pulses are applied by means of the radiofrequency transmitter coil which is located within the magnet, close to, or surrounding, the sample. The coil forms part of an electrical circuit which is tuned to the resonance frequency and is therefore capable of delivering radiofrequency fields of the requisite strength. This circuit is essentially the final stage of the transmitter.

The signal that is observed in response to a radiofrequency pulse or a series of pulses does not look like a spectrum or an image; in its simplest form it may appear as an oscillation which gradually decays away, as shown in Fig. 1.13. This signal is referred to as a *free induction decay* (FID), and it decays to zero with the time constant T_2^*. More generally, a free induction decay contains many frequency components; it then has a much more complex appearance, and is very difficult to interpret visually. In order to transform

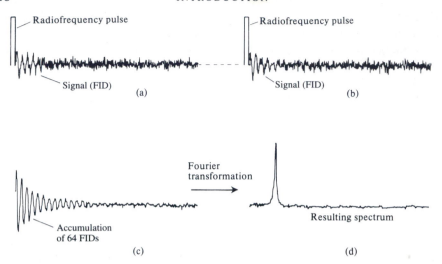

Fig. 1.13 Detection of signals using Fourier-transform NMR. This diagram illustrates the accumulation of a spectrum containing a single resonance. (a) and (b) show two consecutive free induction decays acquired in response to 90° radiofrequency pulses. (c) is the accumulation of 64 free induction decays (scaled down from (a) and (b) by a factor of 16) and (d) is the spectrum obtained on Fourier transformation of the accumulated signal. (From Gadian 1977.)

a free induction decay (or a series of such decays) into an understandable form, it is necessary to apply a mathematical manipulation known as Fourier transformation. Essentially, this converts a time-dependent signal into its equivalent frequency components, thereby generating a recognizable spectrum or image. If we consider a spectrum containing several resonances, then the linewidths of the various resonances will be inversely related to the corresponding T_2^* values, according to the relationship given in eqn (1.10).

It was noted in the above section that NMR has inherently low sensitivity and that this imposes limitations on the concentrations of compounds that can be detected and on the spatial resolution that can be achieved. In practice, the weakness of the signals from metabolites also means that repeated data acquisitions are normally required in order to build up an adequate signal-to-noise ratio. The accumulation of N consecutive acquisitions leads to an improvement of $\sqrt{N}$, because the signal increases by a factor of N whereas the background noise, being random, increases by $\sqrt{N}$. So in Fourier-transform NMR spectroscopy, a radiofrequency pulse is applied, the signal is acquired, and the process is repeated a number of times at intervals of, typically, a second (see Fig. 1.13). The data are automatically added until the required signal-to-noise ratio is obtained, and then Fourier-transformed to produce the final spectrum.

Most imaging techniques also require the use of repeated data acquisitions. This is primarily because it is necessary to collect signals using a range of different magnetic field gradient conditions in order to obtain the required spatial information. However, there is an imaging approach, known as echo-planar imaging, in which a full data set can be acquired in a single acquisition. This has obvious value in situations where 'snapshot' images are desirable, e.g. for imaging moving objects (particularly if the motion is aperiodic) or for following rapid responses to perturbations.

1.10 PULSE SEQUENCES

One of the most remarkable features of magnetic resonance is the extensive range of pulse sequences that have been (and continue to be) developed, with a view to enhancing the quality and information content of spectra or images. For example, innovative pulse sequences have contributed in many ways to improvements in image contrast, spectral localization, suppression of unwanted signals, and visualization of specific structural, biochemical, or functional properties. A number of these sequences, which involve appropriate combinations of radiofrequency pulses and of magnetic field gradients, are described in Chapter 8. Here, we briefly discuss two basic sequences that illustrate the generation of T_1- and T_2-weighted contrast. These sequences exploit 90° and 180° pulses, which have the effect of tilting the magnetization through angles of 90° and 180°, respectively.

The simple pulse sequence shown in Fig. 1.13 forms a basis for giving T_1-weighted contrast. If the time interval TR between the consecutive 90° radiofrequency pulses is insufficient to allow complete T_1 relaxation, then the magnetization is unable to recover fully between pulses, and the signal intensity will therefore be reduced. This reduction in signal intensity is termed saturation, and varies according to the value of T_1 relative to TR. Signals with differing T_1 values will therefore be reduced in intensity to differing extents, giving rise to T_1-weighted contrast. This contrast can be manipulated by varying TR, or alternatively by varying the pulse angle; for example, pulse angles of less than 90° are commonly employed in the acquisition of both images and spectra, as discussed in Section 7.3.

T_2-weighted contrast has its basis in the formation of spin echoes. As the name implies, an echo refers to a signal that is formed, or regenerated, some time after the initial response. During this echo time (or echo delay, TE), the effects of spin–spin relaxation cause the signal to decay with a time constant T_2, and so the signal intensities in spectra and images will be weighted according to the T_2 values of the species that give rise to them. A simple spin-echo sequence is illustrated in Fig. 1.14. The echo is produced by the introduction of a refocusing (180°) radiofrequency pulse applied at a time $TE/2$ after the excitation pulse. The explanation for this echo formation is given in Section 8.3. Another type of echo, termed a gradient echo, can be generated by means of reversal of a pulsed field gradient, rather than by the introduction of a 180° pulse. Gradient echoes are commonly used in rapid imaging sequences; they

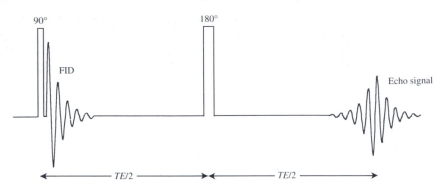

Fig. 1.14 A spin-echo sequence. The sequence consists of a 90° pulse, followed after a time *TE*/2 by a 180° pulse. The effect of this pulse sequence is to produce an echo signal at a time *TE* after the initial 90° pulse.

generate T_2^*-weighted images, and their T_2^*-dependence is proving to be of particular value for functional neuroimaging. More generally, echo signals are used extensively in imaging and spectroscopy, and the behaviour of the magnetization during the echo time *TE* can lead to a myriad of interesting effects and dependencies.

Images with T_1- or T_2-weighted contrast can be obtained by combining simple radiofrequency pulse sequences such as those shown in Figs 1.13 and 1.14 with magnetic field gradients that generate the required spatial encoding. As an illustration of the effects of T_1- and T_2-weighted contrast, Fig. 1.15 shows images of the brain obtained with two different pulse sequences. It can be seen that there is differing contrast between grey and white matter in the two images; the white matter gives higher intensity than grey matter in the T_1-weighted image (Fig. 1.15(a)), but lower intensity in the T_2-weighted image (Fig. 1.15(b)). The contrast arises primarily because the water protons of white matter have shorter T_1 and T_2 values than those of grey matter. The shorter T_1 causes the white matter protons to undergo less saturation, and hence to generate higher signal intensity in the T_1-weighted image. However, because of their shorter T_2 value, the white matter protons undergo a greater degree of signal loss during the echo time, *TE*, of the T_2-weighted sequence.

1.11 APPLICATIONS OF NMR TO BIOLOGICAL SYSTEMS

NMR can be used in a wide variety of ways, but for the investigation of biological systems it will be convenient for us to distinguish between three types of application.

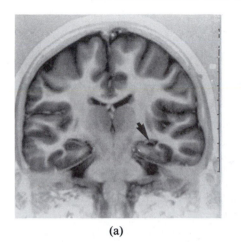 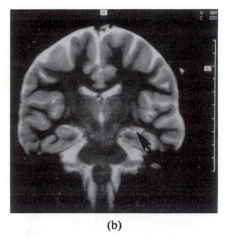

(a)   (b)

Fig. 1.15 T_1-weighted (a) and T_2-weighted (b) images of the brain, illustrating the differing contrast effects generated by the two types of imaging sequence. The arrows refer to the left hippocampus, which in this patient with epilepsy shows abnormalities on both T_1- and T_2-weighted images. (From Great Ormond Street Hospital.)

1.11.1 Structure and function of macromolecules

Since the early 1950s, NMR has been recognized as a powerful means of studying the structures of molecules in solution. Increasingly sophisticated NMR methods have been devised for structural analysis, generally involving the collection of two- (or more) dimensional data sets which provide a great deal of information, not only through chemical shifts and spin–spin couplings, but also through additional connectivities between different nuclei, such as through-space interactions whereby one nucleus influences the relaxation of another, neighbouring nucleus. These data sets can provide detailed information about the structures of molecules in solution, and can be used for monitoring conformational changes resulting from chemical modification, binding of ligands, etc. (Roberts 1993).

While such studies have many technical features in common with NMR of living systems, in many respects the two approaches have diverged as disciplines, and so in this book, which is concerned with living systems, we shall not review the important contribution that NMR has made to our understanding of molecular structure and its relationship to function.

1.11.2 Metabolism

An alternative, more empirical, approach is to use NMR as a means of identifying the presence of particular molecules within a sample. This can be regarded as a 'molecular fingerprint' approach, for it relies on the empirical observation that different molecules produce their own characteristic spectra.

Once the molecules have been identified, any changes in the spectral lines can be interpreted in terms of metabolic changes taking place within the sample. In addition, the precise nature of the spectra can yield further information about the chemical environment of the molecules. This approach can be used for the analysis of body fluids, cell cultures, isolated tissues, small animals, and humans, as discussed in detail in this book.

1.11.3 Imaging

The third approach involves the use of magnetic field gradients to generate images. Much of the emphasis in magnetic resonance imaging has naturally focused on its use in diagnostic radiology, which at first sight has relatively little overlap with metabolic studies. However, as the technology matures it is apparent that there is indeed considerable overlap between imaging and metabolic studies, in terms of the techniques as well as their applications. While service diagnostic radiology will no doubt remain the most widespread application of magnetic resonance imaging, in this book we place more emphasis on the research aspects of the technique.

1.12 NMR STUDIES OF LIVING SYSTEMS—A HISTORICAL PERSPECTIVE

The idea of using NMR to study living systems is by no means new. In fact, very soon after the first successful NMR experiments on bulk matter (Bloch *et al.* 1946; Purcell *et al.* 1946), Bloch obtained a strong proton signal on placing his finger in the radiofrequency coil of his spectrometer (see Andrew 1980). Four years later Shaw and Elsken (1950) used ^{1}H NMR to investigate the water content of potato and maple wood. On the basis of their observations, they suggested that NMR might provide a useful method for the rapid determination of the water content of hygroscopic materials. Odeblad and Lindstrom (1955) obtained low-resolution ^{1}H NMR signals from a number of mammalian preparations, including human red blood cells and striated muscle and fat tissue of the rat. Singer (1959) used ^{1}H NMR to measure blood flow in the tails of mice, and he suggested the possibility of making similar measurements of blood flow in human beings.

In the 1962 edition of his book *Physical Chemistry*, Moore posed the prophetic question:

Suppose a biochemist friend who professed to be quite ignorant about NMR comes to you for advice. He would like to study the transformation ATP → ADP which occurs in muscle metabolism by monitoring the NMR of the ^{31}P nuclei. He is rather wealthy, and has a complete NMR apparatus, including oscillators at 40, 30, 20, 10, and 3 Mc/sec, and a magnet which will go up to about 12000 gauss. Which oscillator would you suggest he use to study the ^{31}P spectra and why? Calculate the value of the magnetic field at which the resonances would appear.

Clearly, at least some of the potential of NMR in biology had been appreciated for many years, but unfortunately the early experiments were limited in scope by the relatively poor instrumentation available at the time. For example, the first 'high-resolution' ^{1}H NMR spectrum of a protein had been reported by Saunders et al. (1957) who were only able to resolve four broad peaks from the enzyme ribonuclease A.

The development of high-field superconducting magnets in the late 1960s, together with the emergence of Fourier-transform NMR, revolutionized the scope of NMR and made a rapid impact on the study of proteins and other biological molecules in solution. However, it emerged rather more slowly that NMR might have extensive applications in the study of metabolites within living systems. Then Moon and Richards (1973) reported high-resolution ^{31}P NMR studies of intact red blood cells. They detected signals from 2,3-diphosphoglycerate, inorganic phosphate, and ATP, and showed how the spectra could be used to determine intracellular pH. The following year, ^{31}P NMR spectroscopy was used for the detection of metabolites in intact, freshly excised rat leg muscle (Hoult et al. 1974). Signals could be assigned to ATP, phosphocreatine (PCr), and inorganic phosphate (P_i), and as in the red cell studies, the chemical shift of the inorganic phosphate signal was used for the measurement of the intracellular pH (Fig. 1.16). In these initial experiments, the muscles were not maintained under controlled physiological conditions. Therefore, as expected, there was a gradual breakdown of high-energy phosphates (a loss of phosphocreatine followed by depletion of ATP), together with a fall in intracellular pH resulting from the accumulation of lactic acid. Similar observations were reported by Burt et al. (1976) for other muscle types. These early studies set the scene for using ^{31}P MRS as a non-invasive probe of tissue energy metabolism.

At about the same time, it was also shown that ^{13}C NMR could be used to follow the end products of metabolic pathways (Eakin et al. 1972; Sequin and Scott 1974). Somewhat later, the use of ^{1}H NMR for metabolic studies was described by Brown et al. (1977), who showed that ^{1}H signals could be observed from a range of compounds in suspensions of red cells, including glucose, lactate, pyruvate, alanine, and creatine. During the mid- to late-1970s, an extensive range of cell types and tissues were investigated, primarily using ^{31}P NMR, but also using ^{13}C NMR (see Cohen and Shulman 1980). Emphasis was placed on the non-trivial matter of maintaining preparations under controlled physiological conditions within the NMR spectrometer, one example being a series of studies on isolated frog muscles, in which it proved possible to stimulate the muscles electrically, measure their tension response, and follow the resulting changes in their ^{31}P NMR spectra (see Dawson et al. 1980). In this way, it was possible to investigate the metabolic basis of muscular fatigue.

In 1980 it was reported that, by placing an unusual type of radiofrequency coil (termed a surface coil) adjacent to a region of interest, it was possible to study skeletal muscle and brain metabolism non-invasively in small animals (Ackerman et al. 1980). Extension to human studies, firstly of limbs (Ross et al. 1981) and then of the neonatal brain (Cady et al. 1983; Younkin et al.

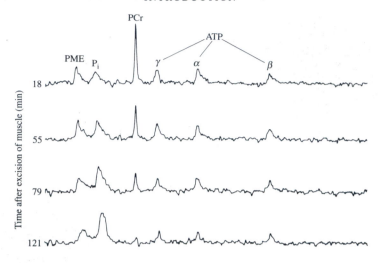

Fig. 1.16 ^{31}P NMR spectra of an intact muscle from the hind leg of a rat. The signals are assigned to the α-, β-, and γ-phosphates of ATP, phosphocreatine (PCr), inorganic phosphate (P$_i$), and phosphomonoesters (PME). (Adapted from Hoult *et al.* (1974), *Nature*, **252**, 285–7 with permission from Macmillan Magazines Limited.)

1984) followed shortly afterwards with the construction of larger magnets. For example, Ross *et al.* (1981) showed a series of ^{31}P spectra obtained from fore-arm muscle before, during, and after exercise, and demonstrated abnormal metabolism in a patient with a disorder of glycogen metabolism. Whole-body systems suitable for both imaging and spectroscopy became available and, as discussed below, the degree of overlap between these two NMR approaches became increasingly apparent.

Imaging developments, based largely on the detection of ^{1}H signals from water, had been proceeding from the early 1970s, but more or less independently of the above developments in spectroscopy. Damadian (1971) reported that certain malignant tumours of rats differed from normal tissues in their ^{1}H NMR properties, and suggested that ^{1}H NMR might therefore have diagnostic value. Lauterbur (1973) published the first NMR image of a heterogeneous object, coining the now obsolete word zeugmatography for this new method of imaging (Fig. 1.17). His method involved applying magnetic field gradients along a number of different directions, obtaining an equivalent number of one-dimensional profiles of the object, and reconstructing a two-dimensional image in a similar manner to the projection–reconstruction procedures used in X-ray scanning (Hounsfield 1973). At about this time field gradients were also being used to investigate periodic structures, in a form of NMR diffraction (Mansfield and Grannell 1973). In the years that followed, a range of NMR imaging techniques were proposed, some of which remain in use, others becoming little more than historical curiosities. An image of a finger was reported in 1976 (Mansfield and Maudsley 1976), studies of the

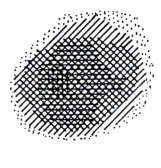

Fig. 1.17 ^{1}H NMR 'zeugmatogram', or image, of two tubes of water. (Reproduced with permission of Macmillan Magazines Limited from Lauterbur (1973), *Nature*, **242**, 190–1.)

hand (Andrew *et al.* 1977) and wrist (Hinshaw *et al.* 1977; see Fig. 1.18) soon followed, and before long whole-body images were being obtained. Image quality continually improved, partly through engineering and technological developments, and partly through increasing skills and experience in the manipulation of field gradients and radiofrequency pulse sequences. By 1980, the clinical evaluation of magnetic resonance imaging had begun, and since that time there have been continuing developments in instrumentation and applications which have led to the present situation in which over 6000 whole-body systems are installed world-wide.

Much of the emphasis in the studies described above was on the use of ^{31}P MRS for the investigation of tissue energetics, and on ^{1}H MRI for structural imaging. Therefore, while it was appreciated that spectroscopy and imaging had a great deal in common, there was nevertheless a tendency, up to the early 1980s, to consider *in vivo* spectroscopy and imaging as two somewhat distinct developments, using different nuclei (^{31}P or ^{13}C for MRS, ^{1}H for MRI) and different instrumentation (high, homogeneous fields for MRS; lower, less homogeneous fields for MRI), with differing aims (MRS being concerned with tissue biochemistry, MRI with applications in diagnostic radiology). However, numerous advances took place from the early 1980s, which tended to reveal the common features of MRI and MRS rather than their differences.

One of the developments was the emergence of ^{1}H MRS for the investigation

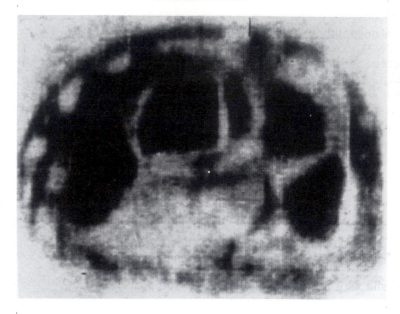

Fig. 1.18 ¹H NMR image of the distribution of mobile protons in a thin transverse section through the wrist. (Reproduced with permission of Macmillan Magazines Limited from Hinshaw *et al*. (1977), *Nature*, **270**, 722–3.)

of brain metabolism. ¹H MRS of metabolites *in vivo* presented technical difficulties because the ¹H signal from water is very much larger than, and can therefore interfere with, the ¹H signals from the metabolites of interest. In addition, the narrow chemical shift range of ¹H spectra imposes particularly severe demands on field homogeneity. However, Behar *et al*. (1983) showed that ¹H spectra of high quality could be obtained from the rat brain *in vivo*, signals being observed from a number of metabolites including *N*-acetylaspartate, creatine + phosphocreatine, and choline-containing compounds. Furthermore, they showed an increase in the lactate signal during hypoxia, with a subsequent return to normal on reoxygenation. While ¹H MRS of tissues other than the brain still presents problems, its applications in the brain have grown rapidly, to the extent that the majority of MRS studies of brain metabolism in humans now use the ¹H nucleus. The fact that these studies and conventional MRI use the same nucleus facilitates the integration of spectroscopy with imaging.

Several additional technical factors brought imaging and spectroscopy closer together. For example, it was shown that images and spectra of the human head could be obtained on the same NMR instrument (Bottomley *et al*. 1983). In addition, surface coils, which provided a simple and sensitive method of obtaining spectra from superficial regions, were found to be useful for imaging too. Pulsed magnetic field gradients came to be used not only for imaging, but also for obtaining spectra from well-defined localized volumes within the body. Moreover, the selection of these volumes of interest, and the interpretation of clinical spectra, commonly relies upon the anatomical detail provided by MRI. Now, it is possible to use spectroscopic imaging methods (Brown *et al.* 1982; Maudsley *et al.* 1983), to obtain images (albeit of relatively crude spatial resolution) that reflect the spatial distribution of metabolites within a tissue.

Further links between spectroscopy and imaging evolved through an increasing awareness of their common goals, not just at the technical level, but also at the level of understanding tissue physiology and pathology. For example, in addition to providing exquisite anatomical detail, major developments have taken place in the use of MRI for visualizing blood vessels and monitoring various aspects of haemodynamics, including blood flow and the effects of blood oxygenation on image intensities. This has led to the emergence of MRI as a new method of mapping brain function, as discussed in detail in Section 4.6. Such studies of the brain provide an indication of the ways in which modern magnetic resonance techniques can provide a non-invasive, integrated approach to the investigation of tissue structure, biochemistry, and function. The following chapters deal in considerably more detail with the physical principles and techniques underlying such observations, and with a range of illustrative applications of MRI and MRS.

REFERENCES

Ackerman, J. J. H., Grove, T. H., Wong, G. G., Gadian, D. G., and Radda, G. K. (1980). Mapping of metabolites in whole animals by ^{31}P NMR using surface coils. *Nature*, **283**, 167–70.

Andrew, E. R. (1980). N.m.r. imaging of intact biological systems. *Phil. Trans. R. Soc. Lond. B.*, **289**, 471–81.

Andrew, E. R., Bottomley, P. A., Hinshaw, W. S., Holland, G. N., Moore, W. S., and Simoraj, C. (1977). NMR images by the multiple sensitive point method: application to larger biological systems. *Phys. Med. Biol.*, **22**, 971–4.

Behar, K. L., den Hollander, J. A., Stromski, M. E., Ogino, T., Shulman, R. G., Petroff, O. A. C., *et al.* (1983). High resolution ^{1}H NMR study of cerebral hypoxia *in vivo*. *Proc. Natl. Acad. Sci. USA*, **80**, 4945–8.

Bloch, F., Hansen, W. W., and Packard, M. (1946). The nuclear induction experiment. *Phys. Rev.*, **70**, 474–85.

Bottomley, P. A., Hart, H. R., Edelstein, W. A., Schenck, J. F., Smith, L. S., Leue, W. M., *et al.* (1983). NMR imaging/spectroscopy system to study both anatomy and metabolism. *Lancet*, **ii**, 273–4.

Brown, F. F., Campbell, I. D., Kuchel, P. W., and Rabenstein, D. C. (1977). Human erythrocyte metabolism studies by ^{1}H spin-echo NMR. *FEBS Lett.* **82**, 12–16.

Brown, T. R., Kincaid, B. M., and Ugurbil, K. (1982). NMR chemical shift imaging in three dimensions. *Proc. Natl. Acad. Sci. USA*, **79**, 3523–6.

Burt, C. T., Glonek, T., and Barany, M. (1976). Analysis of phosphate metabolites, the intracellular pH, and the state of adenosine triphosphate in intact muscle by phosphorus nuclear magnetic resonance. *J. Biol. Chem.*, **251**, 2584–91.

Cady, E. B., Costello, A. M. de L., Dawson, M. J., Delpy, D. T., Hope, P. L., Reynolds, E. O. R., *et al.* (1983). Non-invasive investigation of cerebral metabolism in newborn infants by phosphorus nuclear magnetic resonance spectroscopy. *Lancet*, **i**, 1059–62.

Cohen, S. M. and Shulman, R. G. (1980). ^{13}C n.m.r. studies of gluconeogenesis in rat liver suspensions and perfused mouse livers. *Phil. Trans. R. Soc. Lond. B*, **289**, 407–11.

Damadian, R. (1971). Tumor detection by nuclear magnetic resonance. *Science*, **171**, 1151–3.

Dawson, M. J., Gadian, D. G., and Wilkie, D. R. (1980). Studies of the biochemistry of contracting and relaxing muscle by the use of ^{31}P n.m.r. in conjunction with other techniques. *Phil. Trans. R. Soc. Lond. B.*, **289**, 445–55.

Eakin, R. T., Morgan, L. O., Gregg, C. T., and Matwiyoff, N. A. (1972). Carbon-13 nuclear magnetic resonance spectroscopy of living cells and their metabolism of a specifically labeled ^{13}C substrate. *FEBS Lett.*, **28**, 259–64.

Gadian, D. G. (1977). Nuclear magnetic resonance in living tissue. *Contemp. Phys.* **18**, 351–72.

Hinshaw, W. S., Bottomley, P. A., and Holland, G. N. (1977). Radiographic thin-section image of the human wrist by nuclear magnetic resonance. *Nature*, **270**, 722–3.

Hoult, D. I., Bushy, S. J. W., Gadian, D. G., Radda, G. K., Richards, R. E., and Seeley, P. J. (1974). Observation of tissue metabolites using ^{31}P nuclear magnetic resonance. *Nature*, **252**, 285–7.

Hounsfield, G. N. (1973). Computerized transverse axial scanning (tomography). *Br. J. Radiol.*, **46**, 1016–22.

Lauterbur, P. C. (1973). Image formation by induced local interactions: examples employing nuclear magnetic resonance. *Nature*, **242**, 190–1.

Mansfield, P. and Grannell, P. K. (1973). NMR diffraction in solids? *J. Phys. C: Solid State Phys.*, **6**, 422–6.

Mansfield, P. and Maudsley, A. A. (1976). Planar and line-scan spin imaging by NMR. *Proc. XIXth Congress Ampere, Heidelberg*, 247–52.

Maudsley, A. A., Hilal, S. K. Perman, W. H., and Simon, H. E. (1983). Spatially resolved high resolution spectroscopy by four-dimensional NMR. *J. Magn. Reson.*, **51**, 147–52.

Moon, R. B. and Richards, J. H. (1973). Determination of intracellular pH by ^{31}P magnetic resonance. *J. Biol. Chem.*, **248**, 7276–8.

Moore, W. J. (1962). *Physical chemistry*, Problem No. 36, chapter 14. Longmans Green, Harlow, Essex.

Odeblad, E. and Lindstrom, G. (1955). Some preliminary observations on PMR in biological samples. *Acta Radiol.*, **43**, 469–76.

Purcell, E. M., Torrey, H. C., and Pound, R. V. (1946). Resonance absorption by nuclear magnetic moments in a solid. *Phys. Rev.*, **69**, 37–8.

Roberts, G. C. K. (ed.) (1993). *NMR of macromolecules. A practical approach.* Oxford University Press, Oxford.

Ross, B. D., Radda, G. K., Gadian, D. G., Rocker, G., Esiri, M., and Falconer-Smith, J. (1981). Examination of a case of suspected McArdle's syndrome by ^{31}P nuclear magnetic resonance. *New Eng. J. Med.*, **304**, 1338–42.

Saunders, M., Wishnia, A., and Kirkwood, J. G. (1957). The nuclear magnetic resonance spectrum of ribonuclease. *J. Am. Chem. Soc.*, **79**, 3289–90.

Sequin, U. and Scott, A. I. (1974). Carbon-13 as a label in biosynthetic studies. *Science*, **186**, 101–7.

Shaw, T. M. and Elsken, R. H. (1950). Nuclear magnetic resonance absorption in hygroscopic materials. *J. Chem. Phys.*, **18**, 1113–4.

Singer, J. R. (1959). Blood flow rates by nuclear magnetic resonance measurements. *Science*, **130**, 1652–3.

Younkin, D. P., Delivora-Papadopoulos, M., Leonard, J. C., Subramanian, V. H., Eleff, S., Leigh, J. S., *et al.* (1984). Unique aspects of human newborn cerebral metabolism evaluated with phosphorus nuclear magnetic resonance spectroscopy. *Ann. Neurol.*, **6**, 581–6.

2

The information available from NMR

In this chapter, we describe the type of information that NMR can provide about living systems, concentrating on its role in the study of tissue biochemistry and physiology. If we consider the division of NMR into MRI and MRS, the chapter can be divided into two parts — the information available from the water signal, and the information available from everything else. What is perhaps surprising is how useful the water signal is proving to be, not just for distinguishing different structures within the body, but also for probing normal and abnormal physiology.

One of the most important features of NMR is that it is non-invasive, which is obviously a major advantage for the direct investigation of human subjects. However, it should be emphasized that a great deal can also be learned from the study of body fluids, cell cultures, tissue extracts, isolated tissues, and animal models of disease. Examples of the complementary nature of such investigations will be given.

Another important feature of NMR is that it is non-specific, in the sense that signals may be observed from a large number of compounds, without prior specification of which species should be measured. The simultaneous observation of all of these compounds by MRS, rather than just those selected for chemical analysis, offers the opportunity for detecting unexpected metabolites or metabolic processes which might be missed by more conventional methods.

Whether we are using MRI to obtain structural information or MRS to investigate tissue biochemistry, we first need to consider which compounds give rise to detectable signals. It is not possible here to give an exhaustive list, but some indication can be given by considering in turn the nuclei that are most commonly used for *in vivo* studies. We begin with ^{1}H NMR, which provides the basis of MRI and is also used increasingly for metabolic studies.

2.1 ^{1}H NMR

The proton is the most sensitive nucleus, in that it produces a greater signal-to-noise ratio than any other nucleus (apart from tritium) in a given period of time. The protons of water (and of fats, or lipids) generate strong signals that are exploited in MRI, while the proton signals from other compounds are increasingly used for metabolic studies.

Figure 2.1, which shows ^{1}H and ^{31}P spectra obtained from the human

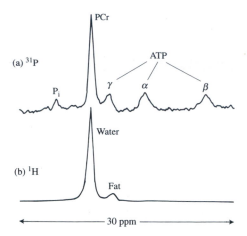

Fig. 2.1 ³¹P and ¹H spectra from the human forearm. The ³¹P spectrum (a) in this early study shows signals from the α-, β-, and γ-phosphates of ATP, phosphocreatine (PCr), and inorganic phosphate (P$_i$). The ¹H spectrum (b) shows signals from water and from fats (lipids). (Adapted from Taylor *et al.* (1983), *Mol. Biol. Med.*, **1**, 77–94, with permission of Academic Press Ltd, London.)

forearm, illustrates a number of important points about the roles of the ¹H and ³¹P nuclei in imaging and spectroscopy. In contrast to the ³¹P spectrum, which provides a direct indication of tissue energy metabolism through the detection of signals from ATP, phosphocreatine, and inorganic phosphate, the ¹H spectrum is relatively featureless and unexciting. It shows a large signal from water, and a somewhat smaller signal from fats. These signals completely dominate the spectrum, because the compounds that give rise to them are present at so much higher concentrations than other species; the concentration of water in tissues is typically about 40 M, or 80 M in protons, whereas the metabolites of interest are in the millimolar range, and this 10 000–100 000-fold difference in concentration translates into an equivalent difference in signal intensities. In practice, this complicates the use of ¹H MRS for the study of tissue metabolites, for it means that the water and fat signals must be suppressed in order to detect the much smaller signals from the metabolites.

On the positive side, however, the use of these very large signals from water and fats enables adequate signal-to-noise ratios to be obtained from very small volume elements. This means that images of very good spatial resolution can be obtained. The spatial resolution that can be achieved for 'metabolite mapping' studies is very much poorer because much larger volume elements are needed in order to obtain adequate signal-to-noise ratios for the metabolites of interest.

The high spatial resolution of proton imaging, together with the contrast

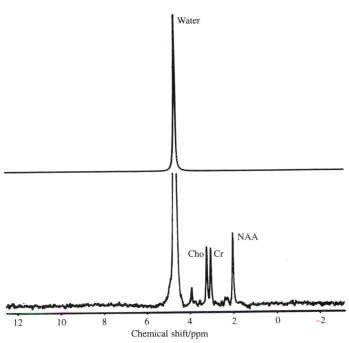

Fig. 2.2 ¹H spectra from the brain (a) with no solvent suppression, showing the dominant signal from water, and (b) with solvent suppression, showing signals from metabolites, with contributions from *N*-acetylaspartate (NAA), creatine + phosphocreatine (Cr), and choline-containing compounds (Cho).

that can be generated through a variety of different mechanisms, leads to the great power of MRI as a method of anatomical imaging and of visualizing pathologies. Numerous properties of the water protons can be used to generate contrast; for example, in addition to the proton concentrations and relaxation times, the effects of diffusion, perfusion, and bulk flow can also be exploited. These effects can also be used as a means of probing physiological processes, as discussed in Section 2.11 below.

The study of tissue metabolites by ¹H MRS is complicated not only by the need to suppress the intense water and fat signals, but also by the large number of metabolites that produce signals in a relatively narrow chemical shift range. Excellent field homogeneity is therefore required, to ensure that the outlying components, or wings, of the water signal do not interfere with the metabolite signals, and also that there is adequate spectral resolution between the signals of interest. It turns out that these various problems are most easily dealt with in the brain, partly because there is little, if any, detectable lipid signal (at least in normal brain), and also because excellent field homogeneity can be achieved. These points are illustrated by Fig. 2.2. Figure 2.2(a)

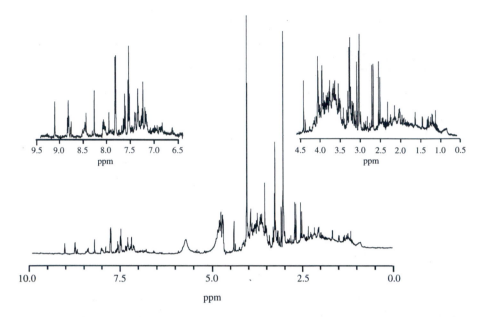

Fig. 2.3 ^{1}H spectrum of urine obtained at 500 MHz. The water signal has been suppressed, and numerous signals can be detected from a wide range of metabolites. The regions between 0.5 and 4.5 ppm, and between 6.5 and 9.5 ppm, are shown on an enlarged scale. Note that the water signal resonates at about 4.7 ppm at 37 °C, but that it has a significant temperature dependence of about 0.01 ppm per °C, resulting in a shift to high frequency (i.e. to the left) as the temperature decreases from 37 °C (see Ackerman *et al.* 1981). As a result, in solution studies that are commonly carried out at room temperature, the water signal typically appears at about 4.9 ppm. (From Preece, unpublished work).

shows a ^{1}H spectrum of the brain, showing just a very narrow signal from the brain water. Figure 2.2(b) shows the spectrum following solvent suppression and enhancement of the vertical scale. Very much smaller signals from brain metabolites can now be seen in the region 2–4 ppm, with dominant contributions from *N*-acetylaspartate, creatine + phosphocreatine, and choline-containing compounds. As discussed in Chapter 3, continued technical developments have permitted an increasingly wide range of additional brain metabolites to be investigated by ^{1}H MRS *in vivo*. In some cases, the detection of specific metabolites of interest requires the use of 'spectral editing' techniques, because of unavoidable problems with spectral overlap; these methods are discussed in Section 8.11.

Much better spectral resolution can be achieved using high-field systems to study relatively small volumes (e.g. 0.5 ml) of body fluids or of cell or tissue extracts (Fig. 2.3). A great deal of information can be derived from such studies. It is evident, for example, that ^{1}H MRS of body fluids has a considerable role

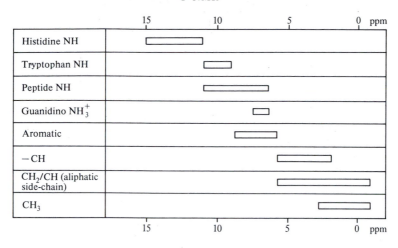

Fig. 2.4 Representative ¹H chemical shift values (measured relative to the standard, tetramethylsilane (TMS)).

to play in areas such as toxicology and the investigation of inborn errors of metabolism (Nicholson and Wilson 1989; Bell 1992); in case of metabolic disease, ¹H spectra of urine samples can provide a rapid means of diagnosis as different metabolic disorders are commonly characterized by unique spectral patterns (Iles and Chalmers 1988).

The general chemical shift ranges that are characteristic of ¹H spectroscopy are given in Fig. 2.4. It is not practical to give a detailed listing of ¹H chemical shifts, but relevant shift values and spectral assignments are given in numerous articles (see, for example, Nicholson and Wilson 1989; Bell 1992; Preece *et al*. 1993).

On the basis of Figs 2.3 and 2.4, it might be anticipated that all ¹H spectra would be displayed with signals to either side of the suppressed water signal. However, spectra obtained *in vivo* commonly display just the region from about 0 to 4 ppm, as the other regions of the spectrum are often featureless or too close to the solvent signal to permit analysis.

2.2 ³¹P NMR

The nucleus that has been used most extensively for metabolic studies is ³¹P, which is the naturally occurring phosphorus nucleus. While ³¹P NMR is less sensitive than ¹H NMR, the chemical shift range is larger (about 30 ppm for biological phosphates; see Fig. 2.5), and it does not have the problem of solvent suppression. Moreover, the dominant signals (see Fig. 2.1(a)) include those of ATP, phosphocreatine (PCr), and inorganic phosphate (P_i), metabolites that play central roles in tissue energetics.

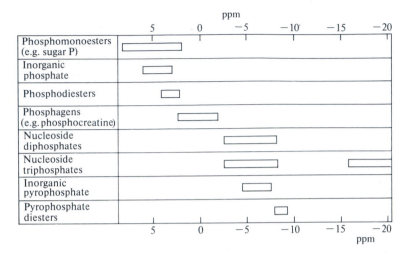

Fig. 2.5 Representative ^{31}P chemical shifts values. (The phosphocreatine signal at pH 7 is assigned the value 0 ppm; see Section 6.1.2 for a discussion of frequency standards.)

The simplicity of the ^{31}P spectrum in Fig. 2.1(a) may appear surprising in view of the large number of phosphorus-containing compounds that are present in muscle. One reason for the relatively small number of peaks is the fact that narrow signals are observed only from relatively mobile compounds; highly immobilized species such as membrane phospholipids normally give very broad signals that are either NMR-invisible or appear as broad components underlying the narrow metabolite signals. Thus, it is commonly assumed that MRS monitors cytoplasmic metabolites (but see Section 2.7 regarding possible contributions from other tissue compartments). Another reason for the simplicity of the spectra is the fact that NMR is an insensitive technique, and therefore detects only those metabolites that are present at fairly high concentrations (typically 0.2–0.5 mM, or greater).

In addition to the ATP, PCr, and P_i peaks, signals are commonly observed *in vivo* from phosphomonoesters and phosphodiesters. Numerous compounds can contribute to these signals, as discussed in Section 2.6, which deals with the question of spectral assignment.

2.3 ^{13}C NMR

Carbon-13 has a natural abundance of only 1.1 per cent, and has a much lower intrinsic sensitivity than ^{1}H NMR. Therefore, in the absence of isotopic enrichment, ^{13}C signals are very weak. As a result, studies using this nucleus can be divided into two categories — those in which compounds are present

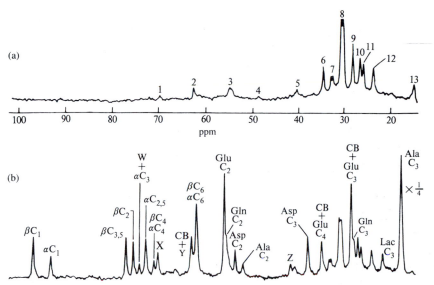

Fig. 2.6 An illustrative ^{13}C study of tissue metabolism. The spectra were obtained at 90 MHz from a perfused mouse liver at 35 °C. (a) ^{13}C natural abundance spectrum obtained before the substrate was added. The peaks labelled 1, 2, 6, 7, 8, 9, 11, 12, and 13 were assigned to the triglycerides of palmitic, oleic, and palmitoleic acids. 8 mM [3-^{13}C] alanine and 20 mM unlabelled ethanol were then added at 0 min and again at 120 min. (b) Spectrum obtained during the period 150–180 min. Peak assignments are as follows: βC_1, αC_1, $\beta C_{3,5}$, βC_2, αC_3, $\alpha C_{2,5}$, αC_4, βC_6, and αC_6 are the carbons of the two glucose anomers; Glu C_2, glutamate; Gln C_2, glutamine; Asp C_2, aspartate; Ala C_2, alanine; Lac C_3, lactate; CB, cell background peak; W, X, Y, and Z, unassigned. (Adapted from Cohen *et al.* 1979.)

at high enough concentrations to be detectable without ^{13}C-labelling for example glycogen in liver and skeletal muscle), and those in which ^{13}C-labelling is used for the investigation of specific metabolic pathways. Such labelling studies have analogies with the use of ^{14}C for radioactive tracer studies, MRS having the advantage that well-resolved signals can be attributed to different carbons within individual molecules. The main disadvantage of MRS is its intrinsic lack of sensitivity, as a result of which the ^{13}C nucleus needs to be present in bulk rather than trace concentrations. In view of the poor sensitivity of ^{13}C relative to ^{1}H NMR, there are advantages in following the ^{13}C nuclei indirectly, by observing protons that are coupled to the ^{13}C nuclei (see Section 8.11). The range of compounds that can be detected is extensive, and is of course influenced by the range of species that can be labelled. This in turn depends not just on chemical considerations, but also on cost; because the ^{13}C nuclei need to be present at high concentrations, the costs of the large quantities of labelled compounds that are required, particularly for human studies, can be very high.

The spectra in Fig. 2.6 provide an illustration of the types of signals that can be seen both in natural abundance and also following administration of ^{13}C-labels. Figure 2.6(a) shows a ^{13}C spectrum obtained from a perfused mouse liver, the various signals being attributed to liver triglycerides that are present at high enough concentrations to give detectable signals in the absence of any labelling. Following the administration of $3\text{-}^{13}C$-labelled alanine, together with unlabelled ethanol, a large number of additional signals can be seen (Fig. 2.6(b)). These can be attributed to a range of metabolites, including glucose, aspartate, glutamate, glutamine, alanine, and lactate. The rate and extent of incorporation of label into the individual carbons of these molecules provides detailed information about the relative rates of many of the reactions of gluconeogenesis.

Figure 2.7 gives representative ^{13}C chemical shift values, and highlights the fact that the ^{13}C nucleus has a wide range of chemical shifts, a feature that helps to compensate for the relatively poor sensitivity of ^{13}C NMR.

2.4 ^{19}F NMR

Fluorine-19 is an excellent NMR nucleus, offering 83 per cent of the sensitivity of ^{1}H NMR. There are no endogenous compounds that give detectable ^{19}F signals *in vivo*, so the applications of ^{19}F NMR involve monitoring exogenous ^{19}F-containing compounds. Examples include the study of blood flow and oxygenation, and pharmacokinetics (in particular of the anti-cancer drug fluorouracil, as discussed in Section 3.4). In addition, fluorinated NMR indicators have been synthesized as intracellular cation probes for H^+ and Ca^{2+}. The use of such indicators is described in Section 2.8.3.

2.5 OTHER NUCLEI

Most of the *in vivo* applications of NMR utilize the nuclei discussed above. Other nuclei have problems of sensitivity, spectral resolution (many nuclei of spin greater than $\frac{1}{2}$ produce very broad lines), or biological relevance. However, there are applications for numerous other nuclei (including ^{14}N, ^{15}N, ^{23}Na, and ^{39}K) in the study of living systems, and some of the ^{23}Na studies are mentioned in Section 4.5.

2.6 IDENTIFICATION OF RESONANCES

The first stage in interpretation of any spectrum must be to identify the molecular groupings that give rise to the observed resonances. Unambiguous identification can sometimes be trivial, but, alternatively, can be extremely difficult, if not impossible. For example, ^{1}H spectra of proteins in solution may contain thousands of resonances, and assignment of these to specific

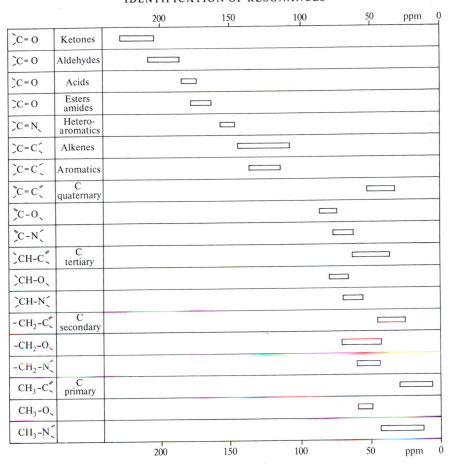

Fig. 2.7 Representative ^{13}C chemical shift values (measured relative to tetramethyl-silane (TMS)).

amino acid residues is an arduous task. In contrast, the ^{31}P spectra of tissues are relatively simple, and assignment of at least some of the signals is straight-forward.

The various methods of assignment are well illustrated by examples from ^{31}P NMR studies. Consider first the ^{31}P spectrum of a perfused rat heart, shown in Fig. 2.8.[1] The signal at about −16 ppm has a chemical shift that is

[1] The phosphocreatine signal that is often present in ^{31}P spectra of intact tissues can provide a suitable and convenient chemical shift reference (see Section 6.1.2). For this reason, and because there are problems associated with using phosphoric acid, which is a commonly used reference, many, but not all, ^{31}P studies of tissue metabolites assign a value of 0 ppm to the phospho-creatine signal. The figures presented in this book use the chemical shift scales used in the original articles from which the figures were obtained.

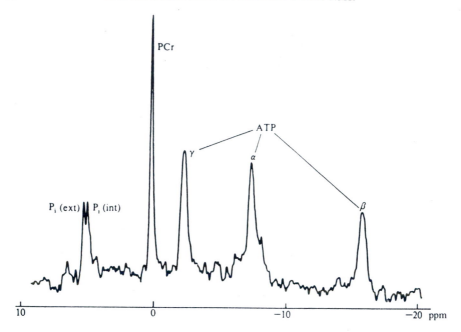

Fig. 2.8 ^{31}P NMR spectrum obtained from a perfused rat heart. The signals are assigned as shown, and as explained in the text. PCr refers to phosphocreatine, and P_i(int) and P_i(ext) to inorganic phosphate in the intracellular space and the perfusion medium, respectively. The chemical shifts are expressed relative to the phosphocreatine signal at 0 ppm (From Grove *et al.* 1980.)

characteristic of the β-phosphate group of nucleoside triphosphates and, on the basis of the known biochemistry of the heart, can be assigned primarily to ATP. (Note, however, that more generally some workers use the label of NTP rather than ATP, in recognition of the fact that nucleoside triphosphates other than ATP can also contribute to the spectra.) The precise value of the chemical shift can be further used to provide information about the intracellular environment of ATP, as discussed in Section 2.8.

Similar reasoning applies for the phosphocreatine and inorganic phosphate signals, both of which can be unambiguously assigned. It may be seen that the inorganic phosphate signal of Fig. 2.8 is split into two components that are shifted slightly from each other. The explanation for this is that inorganic phosphate is present within both the intracellular space and the perfusion medium. These two environments have different pH values, and therefore generate inorganic phosphate signals of slightly different frequencies (see Section 2.7). In order to confirm that the signals do not arise from two different intracellular metabolites, one could make extracts of the heart, and show that these extracts generate just one signal in this region of the spectrum.

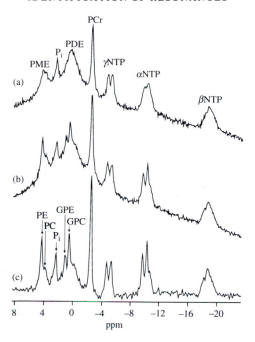

Fig. 2.9 ^{31}P NMR spectra obtained in the course of a chemical-shift imaging study of the human brain. These are non-localized spectra acquired at 1.5 T from the entire head: (a) in the absence of proton decoupling, and (b) in the presence of proton decoupling. (c) shows the data of (b) processed using techniques for removing the underlying broad baseline and enhancing spectral resolution (see Section 7.6.2). (Reproduced by permission of John Wiley and Sons Limited from Murphy-Boesch *et al.* (1993), *NMR Biomed*, **6**, 173–80.)

The signal at −2.5 ppm is characteristic of the γ-phosphate group of ATP, but also overlaps with the signal generated by the β-phosphate of adenosine diphosphate (ADP). Similarly, the signal at −7.5 ppm arises from the α-phosphate group of ATP, but overlaps with the signal from the α-phosphate group of ADP. In addition, a shoulder to the right of the α-phosphate peak can often be seen at a chemical shift characteristic of the phosphate signals of the nicotinamide adenine dinucleotides, NAD$^+$ and NADH. The ADP and ATP signals can be resolved from each other in extracts (see Fig. 2.10, later), but in most spectra of living systems this is not possible, partly because the signals are relatively broad *in vivo*, but also because the contribution from ADP tends to be small.

The other signals commonly observed in ^{31}P spectra are from phosphodiesters (PDE) and phosphomonoesters (PME), as shown in Fig. 2.9(a). Assignment of the PDE and PME signals is not straightforward, because many compounds can contribute to these regions of the spectrum. Following earlier work on skeletal muscle (see Burt 1985 and refs therein), subsequent studies have concentrated on identification of these signals in other tissues, including brain, liver, and tumours. For example, as reviewed by Williams *et al.* (1989), the major contribution to the phosphomonoester signal in brain spectra is believed to be from phosphorylethanolamine. This signal is particularly large in the neonatal brain. However, many other metabolites, including sugar phosphates, can contribute to this region of the spectrum, and in particular could contribute to any changes that occur in disease. The phosphodiester signal contains contributions from glycerophosphorylcholine (GPC) and glyceropho-sphorylethanolamine (GPE), but a relatively mobile fraction of membrane phospholipids can also make a contribution to this region of the spectrum (Cerdan *et al.* 1986). In fact, it appears that there is a field-dependent contribu-tion to tissue spectra from membrane phospholipids, which is greater at 1.5–2 T (the field strengths commonly used for clinical studies) than at the higher fields used for animal and tissue studies (Bates *et al.* 1989; Lowry *et al.* 1992). In the liver, this contribution can be attributed primarily to the endoplasmic reticulum (Murphy *et al.* 1992).

Spectral resolution within the phosphomonoester and phosphodiester regions of the spectrum can be greatly enhanced by using the technique of proton decoupling. As discussed in Section 8.11, this technique can collapse multiplet signals into single lines. The resulting improvement in spectral resolution is apparent in the spectra of Fig. 2.9(b) and Fig. 2.9(c), which now show the presence of individual signals within the phosphomonoester and phosphodiester regions of the spectra. Assignments of these signals can be made on the basis of their chemical shifts and of prior knowledge, although this does not, of course, constitute absolute proof.

Confirmation of spectral assignments is commonly achieved by analysis of tissue or cell extracts. Perchloric acid extracts are generally used for the analysis of water-soluble metabolites, while chloroform–methanol can be used for the extraction of lipid components. NMR spectra of extracts provide additional information because the signals are narrow in comparison with those observed in intact systems. This is exemplified by the ^{31}P spectra of cultured fibroblast cells and their extracts shown in Fig. 2.10; the extract spectrum (Fig. 2.10(b)) clearly shows the multiple components that contribute to many of the signals seen in Fig. 2.10(a). One method of assignment using extracts is to record spectra as a function of pH (see Fig. 2.11). Spectra can be obtained both before and after the addition of expected compounds, and if the additional signal always coincides with the resonance of interest in the extract spectrum, this provides further evidence that the assignment is indeed correct. The procedure of adding compounds to the extract is recommended because the chemical shifts of some signals are sensitive to salt concentration, which is difficult to control in extracts. Further confirmation of assignments can be obtained by adding appropriate enzymes to the extract and observing changes in the spectrum that correspond to enzymatic conversion of one compound to another. Alternatively, standard biochemical procedures can be used for identifying compounds within the extracts.

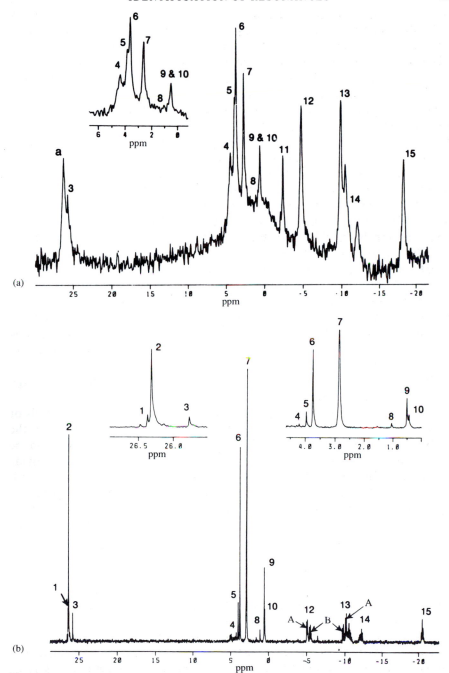

Fig. 2.10 ^{31}P NMR spectra obtained at 162 MHz (a) from perfused fibroblast cells, and (b) from a perchloric acid extract of these cells. The peaks labelled A and B are assigned to NTP and NDP respectively. (Reproduced by permission of John Wiley and Sons Limited, from Swerzgold *et al.* (1994), *NMR Biomed.*, **7**, 121–7, where details of the assignments of the numbered peaks are given.)

While the methods and principles outlined above are also applicable to other nuclei, in some cases the spectral analysis is somewhat more complex. For example, ^{13}C spectra display a range of chemical shift and spin–spin coupling patterns which can be used not just for the identification of individual ^{13}C-labelled carbons within a molecule, but also to establish multiple labelling patterns within molecules, which in turn can be used to provide quantitative information about metabolic pathways and cycles. Specific examples are discussed in Section 3.3.

2.7 pH MEASUREMENTS BY NMR

The methods that have been most commonly used for measuring intracellular pH include (i) insertion of a pH-sensitive microelectrode into the cell, (ii) analysis of the distribution of weak acids or bases, and (iii) colorimetry and fluorometry. These methods all have advantages and drawbacks that are well documented and have been critically reviewed (Roos and Boron 1981).

^{31}P NMR provides an alternative method of measuring intracellular pH, as first shown by Moon and Richards (1973), important points being that the measurements can be made non-invasively and continuously. The method relies on the fact that the frequencies of some of the signals are sensitive to pH and therefore, by means of appropriate calibrations, can be used for the measurement of pH. In practice, the inorganic phosphate signal is most commonly used because it is readily observable in the majority of ^{31}P spectra and because its frequency is particularly sensitive to pH in the physiological range.

Inorganic phosphate (P_i) exists mainly as HPO_4^{2-} and $H_2PO_4^-$ at around neutral pH. In the absence of chemical exchange, these two species would give rise to two signals separated from each other by about 2.4 ppm. In solution, however, the two species exchange with each other very rapidly, and as a result the observed spectrum consists of a single resonance, the frequency of which is determined by the relative amounts of the two species (see Section 6.4). The frequency of the signal measured as a function of pH therefore produces the usual type of pH curve (see Fig. 2.12), and follows the relationship

$$pH = pK_a + \log\left(\frac{\sigma - \sigma_1}{\sigma_2 - \sigma}\right)$$

where pK_a corresponds to the pH at which the $H_2PO_4^-$ and HPO_4^{2-} species are present at equal concentrations, σ_1 and σ_2 represent the chemical shifts of these two species, and σ is the observed chemical shift.

In principle, therefore, it should be possible to determine intracellular pH simply by measuring the chemical shift of the inorganic phosphate signal *in vivo* and determining from the standard titration curve the pH to which this chemical shift corresponds. However, there are a number of potential problems that must be taken into consideration (see Gadian 1983; Petroff *et al.* 1985 and refs therein).

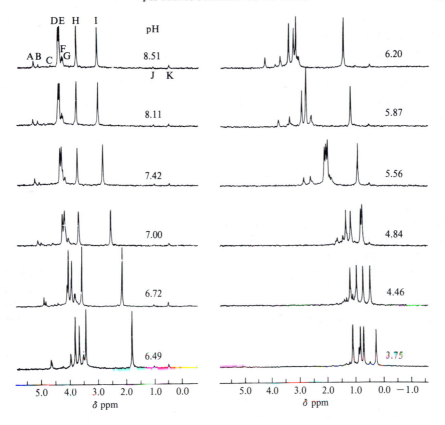

Fig. 2.11 ^{31}P NMR spectra obtained from extracts of tumour cells at 146 MHz showing the pH dependence of signals in the phosphomononester and inorganic phosphate region of the spectrum. The chemical shifts in these spectra are expressed relative to phosphoric acid. (From Navon *et al*. 1977, where details of the assignments are given.)

The first point to consider is whether any factors other than pH can influence the chemical shift of inorganic phosphate *in vivo*; if such effects exist they could generate errors in the estimates of intracellular pH. Certainly, large variations in ionic strength, metal-ion binding, and temperature can influence the chemical shift of inorganic phosphate, largely through their effects on the pK_a of this compound. Therefore, it is advisable to use a titration curve obtained from a solution whose ionic composition resembles that found *in vivo*. Of course, there is bound to be some uncertainty regarding the precise ionic environment within cells, but, fortunately, control experiments such as those shown in Fig. 2.12 have shown that the errors generated by these

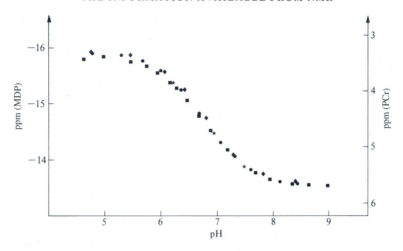

Fig. 2.12 Variation of the chemical shift of inorganic phosphate with pH at 37 °C. The chemical shift is expressed relative to two different standards: (i) methylene diphosphonate (MDP), and (ii) phosphocreatine (PCr) at pH 7. Solutions containing 10 mM inorganic phosphate and 10 mM phosphocreatine were adjusted to various ionic strengths by addition of KCl or NaCl, and titrated at 37 °C by addition of HCl and KOH or NaOH. Symbols: ◆ 120 mM KCl; ■ 160 mM KCl; ● 200 mM KCl; ★ 160 mM NaCl. (Adapted from Garlick *et al*. 19791.)

uncertainties are likely to be small. The binding of inorganic phosphate to macromolecules is another factor that could influence its chemical shift *in vivo*. However, early studies suggested that phosphate–protein interactions had little effect on the titration curves.

An additional problem is that the chemical shift of any signal is measured by comparing its frequency to that of a reference compound. It is therefore essential to have available a suitable reference signal. For ^{31}P NMR, this is provided by phosphocreatine when present; phosphocreatine has a pK_a of 4.6, and therefore its signal is insensitive to pH changes in the normal physiological range. If phosphocreatine is not present, comparison with the ATP signals may be acceptable. Alternatively, the ^{1}H signal from water can provide a useful frequency standard.

A final problem is that there may be some uncertainty as to the distribution of inorganic phosphate within the various tissue compartments. The basis for inorganic phosphate being a monitor of intracellular pH is that most of it is located within the cells, rather than in the extracellular space. However, it should be appreciated that in pathologies where the extracellular space is greatly increased, it is possible that the signal from extracellular P_i might make a significant contribution to the overall signal. The question also arises as to the distribution of inorganic phosphate within the various intracellular compartments. For example, mitochondria occupy about 40 per cent of the intracellular volume of the heart, and so it is conceivable that mitochondrial P_i might contribute to the spectra. A mitochondrial P_i signal has in fact been reported (Garlick et al. 1992), but, nevertheless, it is generally accepted that ^{31}P NMR most commonly reports on the cytoplasmic pH. This view is based on the results of numerous studies. For example, pH measurements using the P_i signal have been shown to agree with measurements from other NMR indicators that are known to be located exclusively in the cytoplasm, and not in mitochondria or the extracellular space. More generally, there is reasonable agreement with other methods of pH determination (see Williams et al. 1989 for discussion). The evidence is now fairly firm that NMR determinations of pH are accurate to within ±0.1 pH units in absolute terms, and that pH changes can commonly be measured with a precision of ±0.03 pH units.

An inorganic phosphate titration curve that is commonly quoted for in vivo studies uses parameters derived by Petroff et al. (1985) for their studies of brain metabolism. Another titration curve, quoted by Taylor et al. (1983) in their studies of muscle metabolism, gives similar results. One caveat concerns the use of these curves for pH measurements of tissue extracts. Perchloric acid extracts may have very high ionic strengths, and as a result pK_a values may be markedly reduced. Titration curves derived for in vivo studies may therefore not be appropriate for analysis of extracts. This is one reason why, as discussed in Section 2.6, it is recommended that assignments based on extract studies should be confirmed by the addition of expected compounds.

The usefulness of ^{31}P NMR for non-invasive measurements of intracellular pH is exemplified by a number of studies described in Chapter 3. These illustrative studies of muscle exercise, ischaemia, and tumour metabolism show how a knowledge of intracellular pH allows a better understanding of metabolic regulation and of tissue energetics.

2.8 MEASUREMENTS OF METAL IONS

Metal ions play essential roles in numerous cellular processes, including the triggering and modulation of metabolic reactions, the maintenance of membrane potentials, and signal transduction. The measurement of intracellular and extracellular metal-ion concentrations is therefore critical to our understanding of these processes.

NMR provides a number of possible approaches to the measurement of metal ions, as reviewed by Ingwall (1992). One approach involves the direct observation of signals from the ions, the most obvious example being the detection of ^{23}Na signals. A second approach, analogous to the use of ^{31}P NMR for the measurement of pH, takes advantage of the binding of metal ions to endogenous NMR-detectable species, such as ATP or citrate; the metal ions influence the NMR properties of the species to which they bind, in much the same way that H^+ ions influence the chemical shift of the inorganic phosphate signal. Similarly, use can be made of signals from exogenous markers that are designed to report on specific metal ions. In this respect, particular interest has focused on ^{19}F-labelled markers.

2.8.1 ^{23}Na NMR

The direct measurement of signals from metal ions is best illustrated by the use of ^{23}Na NMR. Sodium-23 is a nucleus of reasonably high sensitivity, and signals can be readily detected at the typical concentrations found *in vivo*. However, there are two main problems in interpretation of the signals. First, the intracellular and extracellular components have similar chemical shifts. This makes it difficult to distinguish the signals from the two components, and since the extracellular concentration is generally so much higher than the intracellular concentration, the measurement of the intracellular ^{23}Na component is particularly difficult. It is for this reason that paramagnetic 'shift reagents' have been designed for shifting the extracellular signal away from the intracellular signal (Pike and Springer 1982; Gupta and Guipta 1982). The most widely used reagents have been dysprosium chelates such as Dy-TTHA and Dy-DTPA. Unfortunately, the routine application of such reagents *in vivo* is not straightforward. This is partly because a heterogeneous distribution of paramagnetic species also generates local field inhomogeneities (see Section 6.5). While these magnetic susceptibility effects can be specifically exploited for the generation of contrast in MRI (see, for example, Sakuma *et al.* 1994; see also Section 4.4 and 4.6), unfortunately they can obscure the more specific shift effect. It is also necessary to consider possible toxic effects at the concentrations that are required to produce adequate shifts. For these reasons, the search for improved reagents continues. For example, Bansal *et al.* (1993) have reported that relatively low concentrations of the thulium shift reagent Tm-DOTP enables intracellular and extracellular ^{23}Na signals to be resolved *in vivo* (Fig. 2.13), with better spectral resolution of the shifted and unshifted signals than that provided by Dy-TTHA.

The second problem with ^{23}Na NMR is that the ^{23}Na nucleus has a spin of $\frac{3}{2}$ and is therefore quadrupolar. This means that, in comparison with nuclei of spin $\frac{1}{2}$, additional relaxation processes come into play, leading to broader lines, shorter relaxation times, and generally more complex relaxation behaviour. One consequence of this is that the ^{23}Na signal is not always 100 per cent visible; in fact it has been reported to be about 40 per cent visible in many living systems. This 40 per cent visibility is consistent with just one (the $+\frac{1}{2}$ to $-\frac{1}{2}$ transition) of the three spin transitions being detectable, the transitions between the $\frac{3}{2}$ and $\frac{1}{2}$ spin states being too broad to observe (Shporer and Civan 1977). However, the degree of visibility may depend on the tissue under investigation, the physiological conditions, and the NMR

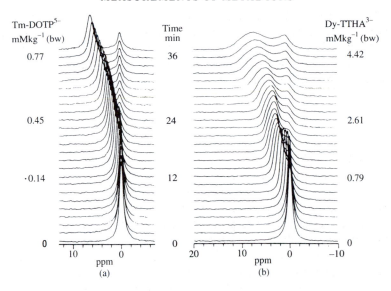

Fig. 2.13 Comparison of ^{23}Na spectra from rat liver during infusion of Tm-DOTP (a) and of Dy-TTHA (b). The time indicated in the middle of the stacked plots is at the beginning of data acquisition. The infusion doses at the corresponding time intervals are shown at the left for Tm-DOTP and at the right for Dy-TTHA. The line-broadening associated with magnetic susceptibility effects is particularly severe with the Dy-TTHA infusion. (From Bansal *et al.* 1993.)

methods that are used for signal detection, which greatly complicates the conversion of signal intensity to absolute concentrations.

As an illustration of these points, Lee *et al.* (1992) have carried out a detailed analysis of the relaxation properties of the ^{23}Na signals of yeast cell preparations. They found several relaxation components, the shortest of which ($T_2 = 1.2$ ms) would be consistent with the broad $\frac{3}{2}$ to $\frac{1}{2}$ spin transition of intracellular Na$^+$. The NMR visibility of such short T_2 components would depend strongly on the characteristics of the pulse sequence and instrumentation used; as a consequence, the apparent visibility of the intracellular ^{23}Na signal in these cells could range from 40 to 100 per cent, depending upon the extent to which the short T_2 components are detected. This could account for some of the variability that is apparent from the literature. This study further suggested that, while the relaxation properties of the intracellular and extracellular components differ from each other, this difference may not be sufficient to allow complete discrimination of the two components unless a rigorous analysis of the relaxation characteristics is carried out or unless a shift reagent is used. In fact, with the use of a shift reagent, Bansal *et al.* (1993) were able to show that the intracellular ^{23}Na signal within the liver has a fast T_2 component of 1.3 ms, which is very similar to the value found in yeast cells. However, these liver data also suggested that there may be additional components that may have even shorter T_2 values and hence remain NMR-invisible.

Such quantitative measurements appear to be clarifying a somewhat confusing area of NMR research, and considerable progress is now being made with the detection of changes in Na^+ distributions, for example following ischaemia, as outlined in Section 4.5.

2.8.2 Measurement of Mg^{2+}

Although ^{25}Mg generates NMR signals, ^{25}Mg NMR is rarely used in living systems, because of the poor sensitivity and broad lines that are characteristic of ^{25}Mg NMR. Instead, measurements of Mg^{2+} *in vivo* rely on the observation of chemical shifts that binding of Mg^{2+} induces in other species. The pioneering work in this area was performed by Cohn and Hughes (1962) who showed that the three ^{31}P resonances of ATP are considerably shifted on binding of divalent metal ions, including Mg^{2+}. Unfortunately, their results are not immediately applicable to studies of living systems because spectrometer sensitivity at that time was such that they had to use 100 mM ATP. At this high concentration, there is considerable stacking of the aromatic rings of the ATP molecules which can affect the chemical shifts. pH titrations of 5 mM ATP in the presence and absence of Mg^{2+} ions are shown in Fig. 2.14. On the basis of these titrations, it is clear that the chemical shifts of the ATP signals observed *in vivo* should provide information about the extent to which ATP is complexed to Mg^{2+} ions in the cell. (In principle, binding could also occur with other divalent metal ions such as Ca^{2+}, but in general Mg^{2+} will be the only ion present at high enough concentrations to allow significant binding.) Generally, it has been found that most of the ATP *in vivo* is complexed to Mg^{2+} ions. More specifically, from a knowledge of the binding constant of Mg^{2+} to ATP, the concentration of free Mg^{2+} ions has been measured in a number of systems, including red cells, skeletal muscle, and normal and diseased brain (Taylor *et al*. 1991; Helpern *et al*. 1993 and refs therein). The main problem with this method is that in some tissues, the concentration of Mg^{2+} is sufficiently high (around 1 mM) to ensure that the ATP is more or less fully complexed to Mg^{2+} ions. Under these circumstances, the chemical shifts are no longer very sensitive to changes in Mg^{2+} ions, making it difficult to perform accurate concentration measurements. The concentration of Mg^{2+} in the perfused liver has also been assessed from its effects on the ^{13}C signals of citrate, the results being consistent with values obtained using the chemical shifts of ATP (Cohen 1983).

All of these measurements are of value because it is Mg-ATP rather than free ATP that is the substrate for most reactions involving ATP. In addition, knowledge of the concentration of free Mg^{2+} is invaluable when studying enzymes such as creatine kinase for which the equilibrium constant is sensitive

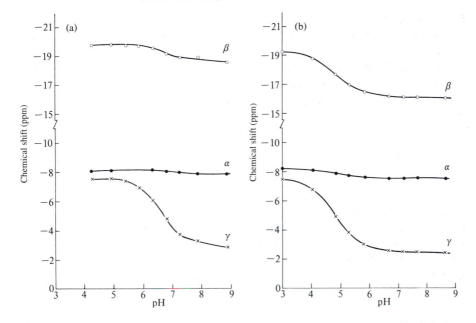

Fig. 2.14 The chemical shifts of the ^{31}P NMR signals of 5 mM ATP plotted as a function of pH, (a) in the absence of Mg^{2+} ions, and (b) in the presence of 10 mM $MgCl_2$, when the ATP is almost fully complexed to Mg^{2+}. (From data presented in Gadian *et al.* 1979.)

to Mg^{2+} concentration. Moreover, many enzymes require Mg^{2+} ions for activity, and therefore their activity could be strongly dependent on the free Mg^{2+} concentrations.

2.8.3 The use of exogenous markers

Ca^{2+} ions play a critical role in the triggering of many biological processes. The intracellular Ca^{2+} concentration is very low (typically about 1 μM), and its measurement has, in the past, exploited the effects of Ca^{2+} ions on fluorescent labels introduced into cells in the form of labelled Ca^{2+}-ligands. As a modification of this approach, analogous ligands have been labelled with fluorine (rather than with fluorescent markers) in such a way that Ca^{2+}-binding influences the ^{19}F NMR signal. Considerable skill with ligand design is required to ensure that the binding of Ca^{2+} to the ligand is such that the ligand is sensitive to changes in Ca^{2+} concentration in the physiological range (i.e. around 1 μM). This innovative approach has been used in the study of isolated cardiac muscle (Smith *et al.* 1983) and of brain slices (Bachelard *et al.* 1988). For example, Fig. 2.15 shows ^{19}F spectra obtained

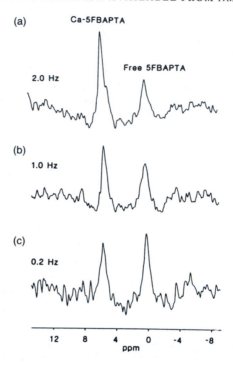

Fig. 2.15 ^{19}F NMR spectra obtained from 5F-BAPTA-loaded hearts at pacing frequencies of 2.0, 1.0 and 0.2 Hz. As the pacing frequency is reduced, the ratio of the bound to free peak is also reduced, indicating a fall in end-diastolic [Ca^{2+}], which was estimated as 1014 nM at 2.0 Hz, 586 nM at 1.0 Hz, and 324 nM at 0.2 Hz. (Reproduced by permission of *Magn. Reson. Med.* from Harding *et al.* (1993), *Magn. Reson. Med.*, **29**, 605–15.)

from a perfused rat heart containing the fluorinated indicator 5F-BAPTA, which had been administered in the form of an ester. Separate signals are observed from Ca^{2+}-bound and free 5F-BAPTA, and the ratio of the two signal areas can be used to determine the intracellular Ca^{2+} concentration.

Unfortunately, as discussed by Harding *et al.* (1993), loading of the heart with the prototype compound 5F-BAPTA at the concentrations required for detection of ^{19}F signals (about 100 μM and above) is associated with a substantial decrease in contractility. This effect on contractility can be attributed to calcium buffering which is also reflected by the relatively high end-diastolic Ca^{2+} concentrations compared to the values obtained using fluorescent indicators. This has prompted work on the development of indicators with enhanced NMR sensitivity, and studies have recently been described using a new high-affinity indicator, DiMe-5F-BAPTA (Kirschenlohr *et al.* 1993). Another approach that is being addressed involves the use of

^{13}C-labelled indicators, which might permit the selective detection of ^{13}C-attached protons using proton-observe carbon-edited spectroscopy (Robitaille and Jiang 1992).

Analogous approaches can also be used in developing exogenous markers for metal ions other than Ca^{2+}. For example, some progress has been made with the development of a ^{19}F NMR indicator, termed FCrown-1, for intracellular Na^+ (Smith *et al.* 1993).

2.9 METABOLITE CONCENTRATIONS AND KINETIC MEASUREMENTS

2.9.1 Concentration measurements

The relative intensities of the signals within a spectrum or image are proportional to the numbers of nuclei that give rise to them. Therefore, relative concentrations of molecules within a sample can be determined from the intensities of their respective signals, but only if appropriate controls have been carried out for the other factors that influence signal intensities. In particular, the effects of relaxation need to be determined; these include signal losses resulting from partial saturation and, in the case of spin-echo studies, from T_2 relaxation. The assessment of these effects is not always straightforward, if only because the measurements can be very time-consuming.

In MRI, images that are acquired in such a way that the signal intensities have no significant dependence on T_1 or T_2 (or on any of the other factors, such as flow, that can influence signal intensities) are termed proton density images. In such images, the relative intensities of different picture elements reflect the relative concentrations of the protons that give rise to them. Thus, oedematous regions or fluids such as cerebrospinal fluid will tend to show up brightly in proton density images because of their relatively high water concentrations in comparison with neighbouring tissue. However, clinical imaging studies rarely use proton density maps to derive *quantitative* information about the distribution of water or about changes in water concentration.

In contrast, spectroscopy relies heavily on quantitative measurements of relative or absolute concentrations of metabolites. In many spectroscopy studies, the measurement of *relative* concentrations, as measured from the relative areas of the respective peaks (see Fig. 1.10), provides an adequate basis for interpretation. In some cases, however, this is not so, and measurements of *absolute* metabolite concentrations are required. However, such measurements are more difficult for a variety of reasons; for example, they require calibration of the signal intensities against that of a known standard, either external or internal.

One general problem of relevance to the measurement of both relative and absolute concentrations is that the precise measurement of signal areas is often difficult. This is partly because the signal-to-noise ratio may not be very high, but also because of spectral overlap between neighbouring peaks within a spectrum. Increasing use is being made of computer fits that give automated analyses of peak areas (de Beer and van Ormondt 1992). However, while the development of increasingly sophisticated software may improve accuracy, save time, and avoid subjectivity, it should be appreciated that peak-fitting routines may rely on assumptions that are not necessarily valid. In particular, it often needs to be assumed that signals have a characteristic lineshape (e.g. Lorentzian or Gaussian). In practice, however, lineshapes can be influenced by many effects, including unresolved spin–spin couplings and magnetic field inhomogeneities, and the resulting uncertainties in the precise lineshape may affect the accuracy of automated procedures for measuring areas, particularly when there is significant spectral overlap. These problems are greatly reduced when the signals are well resolved, and since magnetic field homogeneity strongly influences the resolution of spectral lines, there is no substitute for good 'shimming' (i.e. adjustment of field homogeneity); if the signals are well-resolved, spectra become a lot more informative, as well as being a lot easier to analyse. Even under these circumstances, however, some caution is needed if a high degree of accuracy is required; for example, it may be shown that if a Lorentzian peak is integrated over a frequency range that is as much as ten times the linewidth, this still only gives 94 per cent of the total peak area.

With improvements in spectral quality, localization techiques, and methods of spectral analysis, these difficulties can be more readily addressed, and numerous measurements of absolute concentrations are now emerging (see Cady 1992 for review). For example, several groups have measured the absolute concentrations of metabolites that give rise to the dominant signals in ^{1}H spectra of the brain (see Kreis *et al.* 1993 and refs therein).

2.9.2 Kinetic studies

One of the main advantages of NMR for metabolic studies is that the kinetics of reactions can be followed non-invasively, simply by monitoring how the signal areas vary with time. The scope of this approach is limited, of course, by the time required to obtain satisfactory spectra, which may vary from seconds to hours; however, this time window is well suited for investigations of muscle exercise, and of events associated with ischaemia, as exemplified by studies described in the following chapter. Moreover, in situations where rapid and repetitive changes occur, time resolution can be enhanced by synchronizing NMR data collection with different phases of the cycle.

NMR can also be used to investigate processes that take place under steady-state or equilibrium conditions (for review, see Brindle 1988). Of particular interest for studies of enzyme-catalysed reactions *in vivo* are two different types of 'magnetic labelling' study. In one type of experiment, a magnetic label in the form of a suitable isotope (commonly ^{13}C) is introduced, and its fate monitored through its influence on the spectra. The main drawback is the standard NMR problem of sensitivity and required concentration range for signal detection. Nevertheless, there is scope for labelling and monitoring numerous different metabolites, and this approach has extensive applications, as described in Sections 2.3 and 3.3.

The second type of study uses the NMR technique of magnetization transfer. In such studies, the magnetization of one species is perturbed, and this perturbation can be transferred to a second species if the two are in exchange with each other. Provided that appropriate controls are carried out, the extent to which the second species is affected can give a measure of the rate of interconversion. This method has been widely used to measure the rate of interconversion of phosphocreatine and ATP through the creatine kinase reaction. Furthermore, in some tissues it has been possible to observe magnetization transfer between the ATP and inorganic phosphate signals and hence obtain information about the rate of ATP turnover. Unfortunately, only a limited number of enzyme-catalysed reactions can be studied in this way, partly because they need to be rapid and also because they must involve at least one substrate that gives rise to a detectable signal. However, it should be emphasized that this technique does provide a totally non-invasive means of measuring the steady-state kinetics of certain key reactions as they take place *in vivo*, and some interesting conclusions have emerged, as discussed in Section 3.1. The use of magnetization transfer to determine exchange rates, together with numerous other effects of magnetization transfer, is described in some detail in Section 6.4.

2.10 TISSUE AND CELLULAR HETEROGENEITY

Tissues do not have a homogeneous metabolic or physiological state, and a detailed assessment of images or spectra requires an appreciation of the heterogeneity that may occur at all levels of tissue organization. For example, brain tissue contains a variety of different cell types, including neurones, astrocytes, and oligodendrocytes, resulting in heterogeneity at a microscopic and macroscopic level. Muscle contains different fibre types, which may differ in their response to exercise, while diseased tissue may show considerable heterogeneity; for example, tumours commonly include necrotic as well as viable regions.

Some aspects of this heterogeneity can be visualized directly by MRI. For example, it was recognized early in the development of MRI that this new imaging technique provided excellent soft-matter contrast, one notable example being the contrast between grey and white matter. More recent MRI developments include angiography showing the vasculature, and diffusion-weighted images showing white-matter tracts (see Chapter 4). Spectroscopy, despite its poor spatial resolution, has also demonstrated metabolic heterogeneity, for example in tumours and in exercising skeletal muscle (see Chapter 3). However, if we wish to address heterogeneity at a microscopic level, we need to know more about the biochemistry, physiology, and NMR properties of the different intracellular and extracellular components. This requires, in turn, the study of model systems, including purified cell preparations.

To exemplify this approach, there is increasing interest in the use of ^{1}H signals as specific indicators or 'markers' of individual cell types within the brain. Attention has so far focused on the signal from N-acetylaspartate (NAA). While the function of this compound remains uncertain, it is believed to be present primarily within neurones (see Section 3.2.3), and so a reduction in NAA relative to other metabolites can be interpreted in terms of selective neuronal loss or damage. The validity of this interpretation can be assessed through the analysis of purified cells. It has been found, as expected, that NAA is present in neuronal preparations, and absent from mature astrocytes and oligodendrocytes (see Fig. 2.16). This provides additional confidence in the use of NAA as a neuronal marker in the mature brain. However, NAA was unexpectedly found to be present in a glial precursor cell known as the oligodendrocyte-type 2 astrocyte (O-2A) progenitor. While little is known about the presence of such cells in the human brain, it does raise the possibility that in the developing brain some caution may be needed in using NAA as a specific neuronal marker. These studies aid interpretation of spectra obtained *in vivo*, not only by helping to establish the cellular distribution of NAA, but also because, in turn, they can help to define its metabolic role.

Of course, as the spectra of Fig. 2.16 clearly demonstrate, NAA is just one of the many metabolites that contribute to spectra of cells and their extracts. The detailed analysis of different cell types should therefore contribute much more extensively to our understanding of tissue biochemistry and function *in vivo*. In this respect, there have been a number of important technical developments enabling NMR measurements to be undertaken of cells that can be maintained under controlled well-oxygenated conditions at the high densities required for NMR investigations (Kaplan *et al.* 1992; Gillies *et al.* 1993 and refs therein).

2.11 PERFUSION, DIFFUSION, AND FLOW

As mentioned at the beginning of this chapter, it is remarkable how much information the NMR signal from water can provide, not only about anatomical structure, but also about normal and abnormal physiology. It is the latter aspects that we concentrate on in this book.

The sensitivity of the NMR signal to motion can cause difficulties in MRI scanning, as it generates so-called motion artefacts which can degrade image quality. This is a particular problem for studies of the body, because of breathing, peristalsis, and cardiac motion. The term artefact implies unwanted

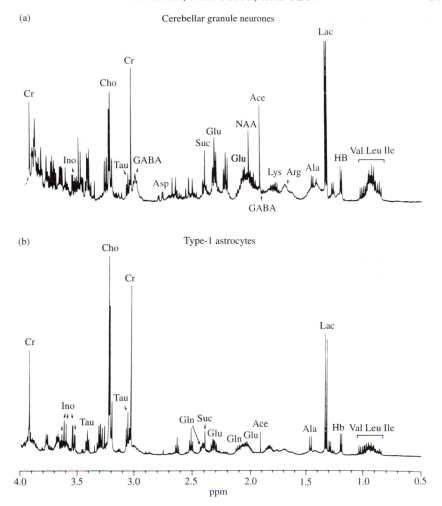

Fig. 2.16 ^{1}H NMR spectra obtained at 500 MHz from extracts of cultured cerebellar granule neurones (a) and cortical astrocytes (b). Identified signals include: valine (Val), leucine (Leu), and isoleucine (Ile); hydroxybutyrate (HB), lactate (Lac), alanine (Ala), lysine (Lys), and arginine (Arg); γ-aminobutyrate (GABA), acetate (Ace), N-acetylaspartate (NAA), glutamate (Glu), glutamine (Gln), succinate (Suc), aspartate (Asp), creatine (Cr), taurine (Tau), choline-containing compounds (Cho), and inositol (Ino). Note that the ^{1}H spectrum of the neurones shows high signals from NAA and the neuroactive amino acids Glu, GABA, and Asp, while the Cho and Cr signals are more dominant in the spectrum of the cortical astrocytes. (Reproduced with permission of the Society For Neuroscience from Urenjak *et al.* (1993) *J. Neurosci.*, **13**, 981–9).

effects, and yet basically the unwanted 'artefacts' and the desired visualization of blood vessels and measurements of flow all arise from the fact that NMR signals are sensitive to motion. Control and manipulation of this sensitivity to motion is leading to improvements in image quality and to the increasing use of NMR as a means of probing tissue physiology. In this section we consider the basic principles underlying the assessment of flow in major vessels, tissue perfusion, and diffusion. Specific applications are discussed in Chapter 4, and details of relevant pulse sequences are given in Chapter 8.

As in other areas of magnetic resonance, the modern MRI approaches to diffusion, perfusion, and flow have built on methods that were originally developed many years ago. For example, it was mentioned in Chapter 1 that Singer (1959) described ^{1}H NMR measurements of blood flow in the tails of mice. He suggested the possibility of making similar measurements in humans, and indeed some years later reported the use of ^{1}H NMR for measurements of venous blood velocities in the human forearm (Morse and Singer 1970). In the 1960s, magnetic resonance methods for measuring diffusion were developed (Stejskal and Tanner 1965; Stejskal 1965). With the aid of modern imaging techniques and technology, these early approaches have been developed and incorporated into imaging protocols so that it is now possible to obtain flow or diffusion-weighted images.

2.11.1 Flow in major vessels

Two guiding principles underlie the sensitivity to flow. The first principle forms the basis of 'time-of-flight' angiography (Fig. 2.17). Suppose that a slab of water is 'magnetically labelled' by the application of radiofrequency pulses. These pulses may, for example, cause the magnetization of the nuclei within the slab to disappear, by the process of saturation (see Section 7.3.2). Under such circumstances, the application of a 90° pulse immediately afterwards would generate no signal from the nuclei within the slab. However, if the 90° pulse were applied after a delay T, then a signal would be generated by the 'unlabelled' spins (shown in black in Fig. 2.17(b)) that have flowed into the slice during the time T. As a result, contrast is generated between stationary and flowing spins, and this provides a means of visualizing blood vessels. In practice, the generation of contrast is complicated by the effects of T_1 relaxation during the time T and by the need for repeated data acquisitions. Nevertheless, this simple example serves to illustrate the basic mechanisms governing the time-of-flight approach to angiography.

The second principle relates to the behaviour of nuclear spins in the presence of a magnetic field gradient. It is shown in Section 8.6 that as spins move through a linear field gradient, their transverse magnetization changes phase at a rate that is proportional to their velocity. NMR methods are available for measuring phase differences, and these can therefore be exploited to display blood vessels and, furthermore, to assess flow rates and velocity distributions within the major vessels.

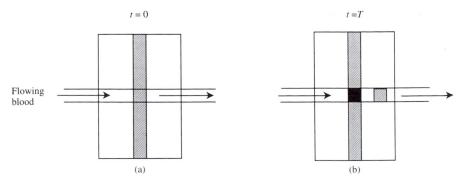

Fig. 2.17 Diagram illustrating the basis of 'time-of-flight' angiography. The rapid application of radiofrequency pulses saturates the spins that are stationary within the image slice (hatched region in (a)). Fully relaxed (unsaturated) spins entering the image slice give greater signal intensity (black region in (b)) than the stationary spins, and the resulting contrast enables angiograms to be generated.

2.11.2 Diffusion and perfusion

Another important effect follows on from the phase changes resulting from motion through field gradients. If there is random motion (Fig. 2.18), then the phases of different spins will vary randomly and, as discussed in Section 8.5, this results in overall loss of signal intensity. While this may appear to be an undesirable effect, there are some important consequences. First, it leads to

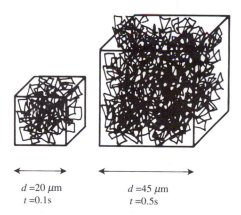

$d = 20 \, \mu\text{m}$ $d = 45 \, \mu\text{m}$
$t = 0.1\text{s}$ $t = 0.5\text{s}$

Fig. 2.18 Diagram illustrating random diffusional motion. d represents the typical distance that a molecule of diffusion coefficient D would diffuse in a time t. Free water would diffuse a distance of about $20 \, \mu\text{m}$ in a time of 0.1s, or $45 \, \mu\text{m}$ in 0.5s. Since $20 \, \mu\text{m}$ approximates to cellular dimensions, it raises the possibility of using diffusion-weighted imaging to probe cellular structure and pathology. (From Le Bihan *et al.* 1991.)

the possibility of measuring diffusion of water, which arises from random Brownian motion. At first sight, the measurement of water diffusion within tissues might appear somewhat uninteresting. However, it turns out that the diffusional properties of tissue water can alter in diseased tissue in such a way that diffusion-weighted imaging is opening up new ways of monitoring and assessing pathology, particularly of the brain (see Section 4.5).

Secondly, the question arises as to whether this approach might provide a method of assessing tissue perfusion, if capillary flow can be modelled in terms of random diffusion-like processes. Obviously, a non-invasive approach to imaging tissue perfusion has major implications, but regrettably, it appears that the difficulties of measuring perfusion by this method are just too severe. However, there are alternative NMR methods which should prove more tractable. A promising approach involves magnetic labelling of water protons by saturation or inversion of arterial spins (Williams *et al.* 1992; Zhang *et al.* 1993). For example, if water protons in the neck region are selectively excited, then arterial flow will carry these magnetically labelled protons into brain tissue and hence influence the observed signal from brain water. Comparison of the data obtained with and without this labelling procedure permits measurements of regional perfusion to be made, as discussed in Section 4.4.

Labelling may also be achieved by the incorporation of suitable isotopes, for example by the use of deuterium-labelled or ^{17}O-labelled water (Ackerman *et al.* 1987; Hopkins *et al.* 1988; Pekar *et al.* 1991). Bolus injections of these labelled species have provided a means of investigating tissue perfusion in animals, but it seems likely that applications in man will be limited, for the labels need to be present at high concentrations, rather than in the trace amounts that are characteristic of radioactive labelling studies. Similarly, there is interest in perfusion measurements exploiting the ^{19}F signal of CHF_3, (Eleff *et al.* 1988; Ewing *et al.* 1990; Branch *et al.* 1992) which can be administered by inhalation rather than by bolus injection. Finally, the use of paramagnetic contrast agents provides an alternative approach to the study of tissue perfusion, as discussed in Section 4.4.

2.12 THE MEASUREMENT OF TISSUE OXYGENATION

There are a variety of NMR approaches to the assessment of tissue oxygenation, some of which are of much wider scope than others. The direct NMR measurement of naturally occurring molecular oxygen is not feasible because ^{16}O has zero nuclear spin and therefore does not give rise to NMR signals. Oxygen-17-labelling studies have been carried out for the assessment of oxygen uptake in animals (Pekar *et al.* 1991) but, as mentioned above in relation to perfusion measurements, their applicability to humans appears limited because of the large amounts of label required. An alternative approach is to take advantage of the fact that molecular oxygen is paramagnetic, and can therefore influence the relaxation properties of adjacent nuclear spins. This could perhaps provide the basis for another method of monitoring oxygenation. In fact, Chiarotti *et al.* (1955) first reported that earlier estimates of 2.3 s for the ^{1}H spin–lattice relaxation time of water contained a contribution from dissolved oxygen, and they obtained a value of 3.6 s on 'degassing' the sample. Unfortunately, the effects of oxygen on the relaxation times of endogenous molecules *in vivo* are likely to be too small to find widespread use. Oxygen does, however, influence the ^{19}F spin–lattice relaxation properties of perfluorocarbons. Perfluorocarbon emulsions have received considerable interest as oxygen-carrying blood substitutes and, by virtue of these relaxation effects, they also provide scope for use as indicators of oxygen tension (Clark *et al.* 1984; Mason *et al.* 1993; Hees and Sotak 1993).

Several other NMR approaches can be used for the assessment of changes in oxygenation state. For example, reduced oxygen delivery to a tissue can lead to marked perturbations in energy metabolites which can be monitored by ^{31}P or ^{1}H NMR, as discussed extensively in Chapter 3. ^{1}H MRS offers opportunities for the detection not only of lactate, but also of other molecules that respond to oxygenation changes, two examples being glutathione and myoglobin.

The ^{1}H NMR spectrum of glutathione depends upon its oxidation state, as was shown by Brown *et al.* (1977) in their studies of red cells. ^{1}H MRS can therefore provide a useful non-invasive measure of the oxidation state of these cells. However, it has proved difficult to extend this approach to other cell systems; for example, in a study of hepatocytes, glutathione was not detected in spectra of whole cells, and gave only weak signals in the spectra of extracts (Nicholson *et al.* 1985). More recently, use has been made of the chemical shift effects induced by the paramagnetic Fe^{2+} centre of deoxymyoglobin. Upon deoxygenation, the electron spin of the haem Fe^{2+} changes from a low spin ($S = 0$) state to a high spin ($S = 2$) paramagnetic state. As a result, ^{1}H signals of deoxymyoglobin are shifted away from the water signal. Of particular interest is the proximal histidyl NH signal, which has a temperature-dependent chemical shift of about 75 ppm when myoglobin is deoxygenated. The concentration of myoglobin in cardiac and skeletal muscle is sufficiently high to permit detection of this signal in hypoxic and ischaemic conditions (Livingston *et al.* 1983; Wang *et al.* 1990; Kreutzer and Jue 1991). In addition, oxymyoglobin is detectable by the ring current-shifted signal from one of its methyl groups (Kreutzer *et al.* 1992). The combined measurement of deoxymyoglobin and oxymyoglobin using this approach could offer an invaluable means of determining cellular oxygenation levels.

Changes in oxygenation state influence the magnetic properties not only of myoglobin, but also of haemoglobin. Thus the Fe^{2+} centres of deoxyhaemoglobin, like those of myoglobin, are paramagnetic, and in addition to perturbing the chemical shifts of signals in the protein spectrum, they also influence the magnetic resonance properties of neighbouring water molecules.

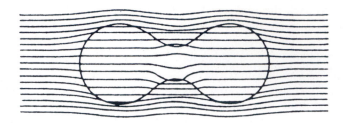

Fig. 2.19 Schematic diagram illustrating how the presence of deoxyhaemoglobin in red cells influences magnetic field lines. The dumb-bell structure represents a red cell containing deoxyhaemoglobin. In the absence of any field gradients, the magnetic field lines illustrated here would be horizontal and of uniform spacing. The distortion of the lines represents the field gradients that are generated by deoxyhameoglobin within the cell. (From Brindle *et al.* 1979.)

The signal intensity in magnetic resonance images can therefore be made to be sensitive to blood oxygenation. Since changes in oxygenation state are commonly associated with physiological and pathological perturbations, images that are sensitized to the oxygenation state of blood provide a powerful non-invasive means of investigating tissue function and dysfunction. As with many other magnetic resonance techniques, attention has focused, at least initially, mainly on the brain.

Deoxyhaemoglobin can influence the water signals through a variety of mechanisms. The main point to emphasize is that its paramagnetic centres generate magnetic field gradients at a microscopic and macroscopic level, and that these cause signal attenuation in T_2^*-weighted images. The spatial characteristics of the field gradients are influenced by the distribution of deoxyhaemoglobin. Thus, the localization of deoxyhaemoglobin to red cells means that there are local field gradients in the periphery of the individual cells (Fig. 2.19), while the overall localization of deoxyhaemoglobin to blood vessels also results, at a more macroscopic level, in the generation of field gradients in the tissue around the vessels. Signal attenuation is therefore not limited specifically to water within the vessels; it can extend significantly into the surrounding tissue. In practice, this means that even if the blood volume is relatively small, there may be substantial signal changes associated with changes in oxygenation state, as is apparent in the brain activation studies that are described in Section 4.6.

REFERENCES

Ackerman, J. J. H., Gadian, D. G., Radda, G. K., and Wong, G. G. (1981). Observation of ^{1}H NMR signals with receiver coils tuned for other nuclides. *J. Magn. Reson.*, **42**, 498–500.

Ackerman, J. J. H., Ewy, C. S., and Shalwitz, R. A. (1987). Deuterium nuclear magnetic resonance measurements of blood flow employing D$_2$O as in inert diffusible tracer. *Proc. Natl. Acad. Sci. USA*, **84**, 4099–102.

Bachelard, H. S., Badar-Goffer, R. S., Brooks, K. J., Dolin, S. J., and Morris, P. G. (1988). Measurement of free intracellular calcium in the brain by ^{19}F-nuclear magnetic resonance spectroscopy. *J. Neurochem.*, **51**, 1311–13.

Bansal, N., Germann, M. J., Seshan, V., Shires, G. T., Malloy, C. R., and Sherry, A. D. (1993). Thulium 1,4,7,10-tetraazacyclododecane-1,4,7,10-tetrakis(methylene phosphonate) as a [23]Na shift reagent for the *in vivo* rat liver. *Biochemistry*, **32**, 5638–43.

Bates, T. E., Williams, S. R., and Gadian, D. G. (1989). Phosphodiesters in the liver: the effect of field strength on the ^{31}P signal. *Magn. Reson. Med.*, **12**, 145–50.

de Beer, R. and van Ormondt, B. (1992). Analysis of NMR data using time domain fitting procedures. In *NMR basic principles and progress* (ed. P. Diehl, E. Fluck, H. Günther, R. Kerfeld, and J. Seelig), Vol. 26, pp. 201–48. Springer-Verlag, Berlin.

Bell, J. D. (1992). MRS of body fluids. In *Magnetic resonance spectroscopy in biology and medicine* (eds J. D. de Certaines, W. M. M. J. Bovee, and F. Podo), pp. 529–57. Pergamon Press, Oxford.

Branch, C. A., Ewing, J. R., Helpern, J. A., Ordidge, R. J., Butt, S., and Welch, K. M. A. (1992). Atraumatic quantitation of cerebral perfusion in cats by ^{19}F magnetic resonance imaging. *Magn. Reson. Med.*, **28**, 39–53.

Brindle, K. M. (1988). NMR methods for measuring enzyme kinetics *in vivo*. *Prog. NMR Spectros.*, **20**, 257–93.

Brindle, K. M., Brown, F. F., Campbell, I. D., Grathwohl, C., and Kuchel, P. W. (1979). Application of spin-echo nuclear magnetic resonance to whole-cell systems. *Biochem. J.*, **180**, 37–44.

Brown, F. F., Campbell, I. D., Kuchel, P. W., and Rabenstein, D. C. (1977). Human erythrocyte metabolism studies by ^{1}H NMR. *FEBS Lett.*, **82**, 12–16.

Burt, C. T. (1985). Phosphodiesters and NMR: a tale of rabbits and chickens. *Trends Biochem. Sci.*, **10**, 404–6.

Cady, E. B. (1992). Determination of absolute concentrations of metabolites from NMR spectra. In *NMR basic principles and progress* (ed. P. Diehl, E. Fluck, H. Günther, R. Kerfeld, and J. Seelig), Vol. 26, pp. 249–81. Springer-Verlag, Berlin.

Cerdan, S., Harihara Subramanian, V., Hilberman, M., Cone, J., Egan, J., Chance, B. *et al.* (1986). ^{31}P NMR detection of mobile dog brain phospholipids. *Magn. Reson. Med.*, **3**, 432–9.

Chiarotti, G., Cristiani, G., and Giuletto, L. (1955). Proton relaxation in pure liquids and in liquids containing oaramagnetic gases in solution. *Nuovo Cim.*, **1**, 863–73.

Clark, L. C. jun., Ackerman, J. L., Thomas, S. R., and Millard, R. W. (1984). High-contrast tissue and blood oxygen imaging based on fluorocarbon ^{19}F NMR relaxation times. *Magn. Reson. Med.*, **1**, 135–6.

Cohen, S. M. (1983). Simultaneous ^{13}C and ^{31}P NMR studies of perfused rat liver. Effects of insulin and glucagon and a ^{13}C NMR assay of free Mg^{2+}. *J. Biol. Chem.*, **258**, 14294–308.

Cohen, S. M., Shulman, R. G., and McLaughlin, A. C. (1979). Effects of ethanol upon alanine metabolism in perfused mouse liver studied by ^{13}C NMR. *Proc. Natl. Acad. Sci. USA*, **76**, 4808–12.

Cohn, M. and Hughes, T. R. (1962). Nuclear magnetic resonance spectra of adenosine di- and triphosphate. II. Effect of complexing with divalent metal ions. *J. Biol. Chem.*, **237**, 176–81.

Eleff, S. M., Schnall, M. D., Ligetti, L., Osbakken, M., Subramanian, H., Chance, B., *et al.* (1988). Concurrent measurements of cerebral blood flow, sodium, lactate and high energy phosphates using ^{19}F, ^{23}Na, ^{1}H and ^{31}P NMR spectroscopy. *Magn. Reson. Med.*, **7**, 412–24.

Ewing, J. R., Branch, C. A., Helpern, J. A., Smith, M. B., Butt, S. M., and Welch, K. M. A. (1990). Cerebral blood flow measured by NMR indicator dilution technique in cats. *Stroke*, **21**, 100–6.

Gadian, D. G. (1983). Whole organ metabolism studied by NMR. *Ann. Rev. Biophys. Bioeng.*, **12**, 69–89.

Gadian, D. G., Radda, G. K., Richards, R. E., and Seeley, P. J. (1979). ^{31}P NMR in living tissue: the road from a promising to an important tool in biology. In *Biological applications of magnetic resonance*, (ed. R. G. Shulman), pp. 463–535. Academic Press, New York.

Garlick, P. B., Radda, G. K. and Seeley, P. J. (1979). Studies of acidosis in the ischaemic heart by phosphorus nuclear magnetic resonance. *Biochem. J.*, **184**, 547–54.

Garlick, P. B., Soboll, S., and Bullock, G. R. (1992). Evidence that mitochondrial phosphate is visible in ^{31}P NMR spectra of isolated, perfused rat hearts. *NMR Biomed.*, **5**, 29–36.

Gillies, R. J., Galons, J-P., McGovern, K. A., Scherer, P. G., Lien, Y.-H., Job, C., *et al.* (1993). Design and applications of NMR-compatible bioreactor circuits for extended perfusion of high-density mammalian cell cultures. *NMR Biomed.*, **6**, 95–104.

Grove, T. H., Ackerman, J. J. H., Radda, G. K., and Bore, P. J. (1980). Analysis of rat heart *in vivo* by phosphorus nuclear magnetic resonance. *Proc. Natl. Acad. Sci. USA*, **77**, 299–302.

Gupta, R. K. and Gupta, P. (1982). Direct observation of resolved resonances from intra- and extracellular sodium-23 ions in NMR studies of intact cells and tissues using DyPPP as a paramagnetic shift reagent. *J. Magn. Reson.*, **47**, 344–50.

Harding, D. P., Smith, G. A., Metcalfe, J. C., Morris, P. G., and Kirschenlohr, H. L. (1993). Resting and end-diastolic $[Ca^{2+}]_i$ measurements in the Langendorff-perfused ferret heart loaded with a ^{19}F NMR indicator. *Magn. Reson. Med.*, **29**, 605–15.

Hees, P. S. and Sotak, C. H. (1993). Assessment of changes in murine tumor oxygenation in response to nicotinamide using ^{19}F relaxometry of a perfluorocarbon emulsion. *Magn. Reson. Med.*, **29**, 311–16.

Helpern, J. A., Vande Linde, A. M. Q., Welch, K. M. A., Levine, S. R., Schultz, L. R., Ordidge, R. J. *et al.* (1993). Acute elevation and recovery of intracellular $[Mg^{2+}]$ following human focal cerebral ischemia. *Neurology*, **43**, 1577–81.

Hopkins, A. L., Haacke, E. M., Tkach, J., Barr, R. G. and Bratton, C. B. (1988). Improved sensitivity of proton MR to oxygen-17 as a contrast agent using fast imaging: detection in brain. *Magn. Reson. Med.*, **7**, 222–9.

Iles, R. A. and Chalmers, R. A. (1988). Nuclear magnetic resonance spectroscopy in the study of inborn errors of metabolism. *Clin. Sci.*, **74**, 1–10.

Ingwall, J. S. (1992). Measuring cation movements across the cell wall using NMR spectroscopy: sodium movements in striated muscle. In *NMR basic principles and progress* (ed. P. Diehl, E. Fluck, H. Günther, R. Kerfeld, and J. Seelig), Vol. 28, pp. 131–160. Springer-Verlag, Berlin.

Kaplan, O., van Zijl, P. C. M., and Cohen, J. S. (1992). NMR studies of metabolism of cells and perfused organs. In *NMR basic principles and progress* (ed. P. Diehl, E. Fluck, H. Günther, R. Kerfeld, and J. Seelig), Vol. 28, pp. 3–54. Springer-Verlag, Berlin.

Kirschenlohr, H. L., Grace, A. A., Clarke, S. D., Shachar-Hill, Y., Metcalfe, J. C., Morris, P. G., *et al.* (1993). Calcium measurements with a new high-affinity n.m.r. indicator in the isolated perfused heart. *Biochem. J.*, **293**, 407–11.

Kreis, R., Ernst, T., and Ross, B. D. (1993). Absolute quantitation of water and metabolites in the human brain. II. Metabolite concentrations. *J. Magn. Reson. B.*, **102**, 9–19.

Kreutzer, U. and Jue, T. (1991). [1]H-nuclear magnetic resonance deoxymyoglobin signal as indicator of intracellular oxygenation in myocardium. *Am. J. Physiol.*, **261**, H2091–7.

Kreutzer, U., Wang, D. S., and Jue, T. (1992). Observing the ^{1}H NMR signal of the myoglobin Val-E11 in myocardium: an index of cellular oxygenation. *Proc. Natl. Acad. Sci. USA*, **89**, 4731–3.

Le Bihan, D., Turner, R., Moonen, C. T. W., and Pekar, J. (1991). Imaging of diffusion and microcirculation with gradient sensitization: design, strategy, and significance. *J. Magn. Reson. Imag.*, **1**, 7–28.

Lee, J-H., Labadie, C., and Springer, C. S. (1992). Relaxographic analysis of [23]Na resonances. *Proc. Soc. Magn. Reson. Med. Berlin Meeting*, Society of Magnetic Resonance in Medicine, Berkeley, CA, p. 2214.

Livingston, D. J., La Mar, G. N., and Brown, W. D. (1983). Myoglobin diffusion in bovine heart muscle. *Science*, **220**, 71–3.

Lowry, M., Porter, D. A., Twelves, C. J., Heasley, P. E., Smith, M. A., and Richards, M. A. (1992). Visibility of phospholipids in ^{31}P NMR spectra of human breast tumours *in vivo*. *NMR Biomed.*, **5**, 37–42.

Mason, R. P., Shukla, H., and Antich, P. P. (1993). *In vivo* oxygen tension and temperature: simultaneous determination using ^{19}F NMR spectroscopy of perfluorocarbon. *Magn. Reson. Med.*, **29**, 303–10.

Moon, R. B. and Richards, J. H. (1973). Determination of intracellular pH by ^{31}P magnetic resonance. *J. Biol. Chem.*, **248**, 7276–8.

Morse, O. C. and Singer, J. R. (1970). Blood velocity measurements in intact subjects. *Science*, **170**, 440–1.

Murphy, E. J., Bates, T. E., Williams, S. R., Watson, T., Brindle, K. M., Rajagopalan, B., *et al.* (1992). Endoplasmic reticulum: the major contributor to the PDE peak in hepatic ^{31}P-NMR spectra at low magnetic field strengths. *Biochim. Biophys. Acta*, **1111**, 51–8.

Murphy-Boesch, J., Stoyanova, R., Srinivasan, R., Willard, T., Vigneron, D., Nelson, S., *et al.* (1993). Proton-decoupled ^{31}P chemical shift imaging of the human brain in normal volunteers. *NMR Biomed.*, **6**, 173–80.

Navon, G., Ogawa, S., Shulman, R. G., and Yamane, T. (1977). [31]P nuclear magnetic resonance studies of Ehrlich ascites tumor cells. *Proc. Natl. Acad. Sci. USA*, **74**, 87–91.

Nicholson, J. K. and Wilson, I. D. (1989). High resolution proton magnetic resonance of biological fluids. *Prog. NMR Spectrosc.*, **21**, 449–501.

Nicholson, J. K., Timbrell, J. A., Bales, J. R., and Sadler, P. J. (1985). A high resolution proton nuclear magnetic resonance approach to the study of hepatocyte and drug metabolism. *Mol. Pharmacol.*, **27**, 634–43.

Pekar, J., Ligeti, L., Ruttner, Z., Lyon, R. C., Sinnwell, T. M., van Gelderen, P., *et al.* (1991). *In vivo* measurement of cerebral oxygen consumption and blood flow using [17]O magnetic resonance imaging. *Magn. Reson. Med.*, **21**, 313–19.

Petroff, O. A. C., Prichard, J. W., Behar, K. L., Alger, J. R., den Hollander, J. A., and Shulman, R. G. (1985). Cerebral intracellular pH by ^{31}P nuclear magnetic resonance spectroscopy. *Neurology*, **35**, 781–8.

Pike, M. M. and Springer, C. S. jun. (1982). Aqueous shift reagents for high-resolution cationic nuclear magnetic resonance. *J. Magn. Reson.*, **46**, 348–53.

Preece, N. E., Baker, D., Butter, C., Gadian, D. G., and Urenjak, J. (1993).

Experimental allergic encephalomyelitis raises betaine levels in the spinal cord of strain 13 guinea-pigs. *NMR Biomed.*, **6**, 194–200.

Robitaille, P.-M. and Jiang, Z. (1992). Improved Ca^{2+} sensitive ligands for *in-vivo* spectroscopy. *Proc. Soc. Magn. Reson. Med. Berlin Meeting*, Society of Magnetic Resonance in Medicine, Berkeley, CA, p. 846.

Roos, A. and Boron, W. F. (1981). Intracellular pH. *Physiol. Rev.*, **61**, 296–434.

Sakuma, H., O'Sullivan, M., Lucas, J., Wendland, M. F., Saeed, M., Dulce, M.C., *et al.* (1994). Effect of magnetic susceptibility contrast medium on myocardial signal intensity with fast gradient-recalled echo and spin-echo MR imaging: initial experience in humans. *Radiology*, **190**, 161–6.

Shporer, M. and Civan, M. M. (1977). The state of water and alkali cations within the intracellular fluids: the contribution of NMR spectroscopy. *Curr. Top. Membr. Transp.*, **9**, 1–69.

Singer, J. R. (1959). Blood flow rates by nuclear magnetic resonance measurements. *Science*, **130**, 1652–3.

Smith, G. A., Hesketh, R. T., Metcalfe, J. C., Feeney, J., and Morris, P. G. (1983). Intracellular calcium measurements by ^{19}F NMR of fluorine-labeled chelators. *Proc. Natl. Acad. Sci. USA*, **80**, 7178–82.

Smith, G. A., Kirschenlohr, H. L., Metcalfe, J. C., and Clarke, S. D. (1993). A new ^{19}F NMR indicator for intracellular sodium. *J. Chem. Soc. Perkin Trans. II*, 1205–9.

Stejskal, E. O. (1965). Use of spin echo in pulsed magnetic-field gradient to study anisotropic, restricted diffusion and flow. *J. Chem. Phys.*, **43**, 3597–603.

Stejskal, E. O. and Tanner, J. E. (1965). Spin diffusion measurements: spin-echoes in the presence of a time-dependent field gradient. *J. Chem. Phys.*, **42**, 288–92.

Swerzgold, B. S., Kappler, F., Moldes, M., Shaller, C., and Brown, T. R. (1994). Characterization of a phosphonium analog of choline as a probe in ^{31}P NMR studies of phospholipid metabolism. *NMR Biomed.*, **7**, 121–7.

Taylor, D. J., Bore, P. J., Styles, P., Gadian, D. G., and Radda, G. K. (1983). Bioenergetics of intact human muscle: a ^{31}P nuclear magnetic resonance study. *Mol. Biol. Med.*, **1**, 77–94.

Taylor, J. S., Vigneron, D. B., Murphy-Boesch, J., Nelson, S. J., Kessler, H. B., Coia, L., *et al.* (1991). Free magnesium levels in normal human brain and brain tumors: ^{31}P chemical-shift imaging measurements at 1.5T. *Proc. Natl. Acad. Sci. USA*, **88**, 6810–14.

Urenjak, J., Williams, S. R., Gadian, D. G., and Noble, M. (1993). Proton nuclear magnetic resonance spectroscopy unambiguously identifies different neural cell types. *J. Neurosci.*, **13**, 981–9.

Wang, Z., Noyszewski, E. A., and Leigh, J. S. jun. (1990). *In vivo* MRS measurement of deoxymyoglobin in human forearms. *Magn. Reson. Med.*, **14**, 562–7.

Williams, S. R., Crockard, H. A., and Gadian, D. G. (1989). Cerebral ischaemia studied by nuclear magnetic resonance. *Cerebrovasc. Brain. Metab. Rev.*, **1**, 91–114.

Williams, D. S., Detre, J. A., Leigh, J. S., and Koretsky, A. P. (1992). Magnetic resonance imaging of perfusion using spin inversion of arterial water. *Proc. Natl. Acad. Sci. USA*, **89**, 212–16.

Zhang, W., Williams, D. S., and Koretsky, A. P. (1993). Measurement of rat brain perfusion by NMR using spin labeling of arterial water: *in vivo* determination of the degree of spin labeling. *Magn. Reson. Med.*, **29**, 416–21.

3

MRS and tissue biochemistry

In this chapter, we discuss the role of magnetic resonance spectroscopy in the investigation of tissue biochemistry. While the focus is on applications of MRS in humans, we also describe some of the ways in which studies of body fluids, cell cultures, tissue extracts, isolated tissue, and animal models of disease can complement and aid interpretation of clinical observations.

It is neither possible nor desirable to give a comprehensive review of this rapidly growing field in a single chapter; instead, we focus on some of the major research themes that are being addressed with MRS. For example, many of the studies described in this chapter relate to the metabolic pathways whereby glucose (or glycogen) is converted by non-oxidative reactions to lactate, or oxidatively to carbon dioxide and water (see Fig. 3.1). The breakdown of glucose generates metabolic fuel in the form of ATP (36 molecules of ATP per molecule of fully oxidized glucose, but just two molecules of ATP if the glucose is broken down as far as lactate). MRS provides a non-invasive means of monitoring metabolites that play central roles in these pathways, measuring the rates at which reactions within the pathways take place *in vivo*, and investigating how these rates are controlled in such a way that metabolic fuel is provided according to demand. These non-invasive studies complement the large amount of information that is available from more traditional invasive approaches to the study of tissue biochemistry. They also offer a number of important advantages, as illustrated by the first of the themes that we consider, namely the use of ^{31}P MRS for the investigation of tissue energetics.

3.1 ^{31}P MRS OF ENERGY METABOLISM

^{31}P MRS studies of skeletal muscle featured prominently in the early demonstrations of NMR as a method of studying tissue metabolism, and in the progression from studies of isolated tissue to studies of humans. There are several reasons for this. First, ^{31}P MRS provides a means of detecting ATP, phosphocreatine, and inorganic phosphate, all of which play key roles in muscle energetics; in particular, the technique provides a non-invasive means of observing metabolic processes associated with the transition from rest to exercise (see Section 3.1.2). Also, for the majority of studies, adequate localization can be achieved by taking advantage of the localizing properties of surface coils, which were developed prior to more sophisticated methods of spectral

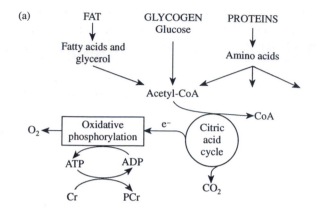

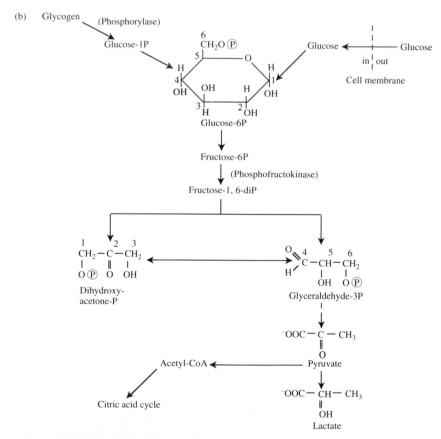

Fig. 3.1 (a) Schematic illustration showing the pathways of oxidative energy metabolism (adapted from Stryer 1988). (b) The glycolytic pathway showing the formation of acetyl-CoA for entry into the citric acid cycle, and the non-oxidative formation of lactate. The individual carbons of the glucose moiety are numbered 1–6, together with the carbons of dihydroxyacetone phosphate and glyceraldehyde 3-phosphate, to indicate the labelling of carbon atoms as they proceed down the pathway into lactate or acetyl Co-A.

localization. In addition, the first magnets that were available for clinical spectroscopy had a bore size that permitted studies of the human forearm, but were not large enough (apart from neonatal studies; see Section 3.5) to accommodate the whole body.

In this section, the role of ^{31}P spectroscopy in the study of energy metabolism is described, primarily in skeletal muscle, but also in other tissues.

3.1.1 Measurement of metabolite levels

Early in the development of ^{31}P spectroscopy, it was naturally important to compare the absolute or relative concentrations of energy metabolites as measured by NMR with those measured more conventionally by the standard invasive approach of freeze-clamping. For frog sartorius muscles, which are thin and can be rapidly frozen, it was shown that freeze-clamping and NMR provide similar values for the concentrations of ATP, phosphocreatine, and inorganic phosphate (Dawson *et al.* 1977). However, in making such comparisons, it should be noted that freeze-clamping estimates usually rely on measurements of *total* amounts of metabolites, whereas NMR generally measures only the mobile components. Therefore, the values obtained by the two techniques could differ, for example, because of tight binding of metabolites to intracellular macromolecules. Values might also differ if there is breakdown of high-energy phosphates associated with the freeze-clamping procedure.

A number of studies have now shown that NMR and freeze-clamping measurements tend to give similar values for tissue ATP concentrations (see Cady 1992 for review). In some cases, however, the two types of measurement may differ somewhat because of the presence of NMR-invisible ATP pools. Also, comparison of the two types of measurement is complicated by the fact that nucleotides other than ATP can make contributions to the triphosphate signals; in recognition of this, these signals are sometimes labelled NTP rather than ATP. In contrast to these ATP observations, the concentrations of ADP and inorganic phosphate (P_i) that have been measured in various tissues using non-invasive NMR methods are often very much lower than those measured by analysis of tissue extracts (Taylor *et al.* 1983; Freeman *et al.* 1983; Iles *et al.* 1985). These discrepancies in the measurements of ADP and P_i could reflect the presence of NMR-invisible intracellular fractions, resulting from tight binding to macromolecules or perhaps from sequestration within certain intracellular compartments. Alternatively, in some cases they may reflect an unavoidable breakdown of high-energy phosphates in invasive measurements, for example, in human biopsy analyses where there is an inevitable delay between removal and freezing of the sample. Regardless of the precise explanation for the differences between these NMR and chemical measurements, the low levels of free ADP and P_i have important kinetic and thermodynamic implications. In particular, they strongly influence our understanding of those reactions that are subject to control by inorganic phosphate and ADP (see Balaban 1990), as exemplified by the studies of human muscle described below.

3.1.2 **Applications to human muscle**

3.1.2.1 *Normal subjects*

NMR is ideally suited to investigating the metabolic changes that are associated with exercise, because the metabolic state can be monitored non-invasively and sequentially throughout a period of rest, exercise, and recovery. While a number of animal studies have been carried out, the advantages of spectroscopy are most apparent for investigations of exercise in humans, where there are obvious benefits over biopsy procedures. Here, a discussion is given of some of the observations that have been made in control subjects. Studies of patients are described in the following section.

Figure 3.2 shows four of a series of ^{31}P spectra obtained from forearm muscle at rest, during aerobic finger-flexing exercise, and during recovery. The spectra were obtained using a surface coil positioned over the finger flexor muscle. Figure 3.2(a) shows a typical resting spectrum, with signals from ATP, phosphocreatine (PCr), and inorganic phosphate. During aerobic exercise, PCr declines, P_i increases (as seen in Figs. 3.2(b) and 3.2(c)) and there is a decline in intracellular pH from the resting value of about 7.0. The decline of PCr rather than of ATP reflects the known capacity of the creatine kinase reaction to maintain the ATP level during exercise (Section 3.1.3). The recovery of PCr following exercise (Fig. 3.2(d)) is normally half completed in about a minute, but is slower when there is significant depletion of ATP during exercise (Taylor *et al.* 1983, 1986).

These NMR observations show broad agreement with previous biopsy studies, with the exception that the resting PCr/P_i ratio is much higher when measured by NMR than by biopsy. Similarly, the concentration of free ADP is very much lower than the total values assessed from biopsy analyses. Other interesting observations include the finding that the intracellular pH can fall to as low as 6.0 on exercise (Taylor *et al.* 1983). In contrast to this, biopsy studies have shown that after exercise to exhaustion, the intracellular pH in the quadriceps muscle falls to about 6.4–6.5. This difference probably reflects the different muscles that were investigated, particularly in view of more recent spectroscopic observations in the human quadriceps muscle which also showed a relatively modest decline in pH, even at the highest workload (Jeneson *et al.* 1992). Whatever the reason for the difference, the NMR results suggest that the glycolytic pathway must be at least partially active at pH 6.0, whereas it had been felt that the pathway might be inactivated at such a low pH, as a result of inactivation of the enzyme phosphofructokinase.

^{31}P MRS offers an opportunity to investigate the mechanisms underlying reduced muscle performance, as in normal ageing and fatigue. In one study, young adult subjects were compared with healthy elderly subjects aged 70–80 years, using ^{31}P spectroscopy to monitor tissue metabolites both at rest and during aerobic, dynamic exercise (Taylor *et al.* 1984). The results suggested

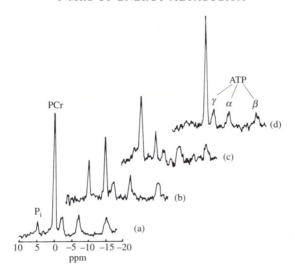

Fig. 3.2 ³¹P spectra obtained from forearm muscle before (a), during (b,c), and after (d) aerobic finger-flexing exercise. Signals are observed from ATP, phosphocreatine (PCr), and inorganic phosphate (P$_i$). (Reproduced with permission from *Magn. Reson. Med.* from Taylor *et al.* (1986), *Magn. Reson. Med.*, **3**, 44–54.)

that the energetics of human skeletal muscle are not altered by the ageing process; the decline in muscular strength and performance may simply be a reflection of reduced muscle mass. ³¹P MRS also provides a useful approach to the investigation of muscular fatigue, as a number of the possible metabolic determinants can be monitored simultaneously in exercising muscle. The problem is that, since a large number of metabolic changes are necessarily associated with exercise, it is difficult to establish which, if any, play a major role in the fatiguing process. One suggestion is that the accumulation of P$_i$ in its monobasic form, $H_2PO_4^-$, is the cause of the decline in force, but this is open to discussion (see Adams *et al.* 1991 and refs therein). Presumably, it is possible that a range of metabolites may be associated with the fatigue process, and that under different circumstances, for example in disease (see below), different factors may be important.

3.1.2.2 *Skeletal muscle studies in disease*
The NMR studies of normal subjects discussed above have provided fresh insights into several different aspects of muscle biochemistry and physiology, but they also give essential baseline information for studies of disease. Abnormalities in tissue energetics have been observed in patients with disorders of oxidative or glycolytic metabolism. For example, in a patient with McArdle's syndrome (in which there is a glycogen phosphorylase deficiency), the muscle

pH became alkaline on exercise, rather than acid (Ross *et al.* 1981), which is totally consistent with the inability of patients with this disorder to generate lactic acid from glycogen. In other early studies, the muscle of a patient with phosphofructokinase deficiency not only became alkaline on exercise, but also accumulated sugar phosphate, as anticipated (Edwards *et al.* 1982). In two sisters with a mitochondrial disorder, recovery of phosphocreatine following exercise was very slow, demonstrating a reduced rate of oxidative metabolism (Radda *et al.* 1982). However, pH recovery was faster than in controls. Furthermore, in view of the high blood-lactate levels that had been observed in one of the patients, the changes in intracellular pH on exercise were somewhat milder than might have been expected, suggesting that such patients might have an adaptive system for eliminating excess acid. Subsequent studies of the mitochondrial myopathies are consistent with these findings, commonly showing one or more of the following: (i) a reduced PCr/P_i ratio at rest; (ii) a small decrease in pH during exercise; and (iii) slow PCr recovery (Matthews *et al.* 1991*a*; Argov and Bank 1991).

It is perhaps not surprising that the most obvious abnormalities in tissue energetics are observed in patients with specific disorders of glycolytic or oxidative metabolism. As discussed in the next section the spectral abnormalities exhibited by these patients are of interest not only in relation to understanding the disease in more detail, but also because they help us to understand the relationships between metabolism and function in healthy muscle.

3.1.3 Control of energy metabolism in skeletal muscle

The transition from rest to exercise involves a large and rapid increase in the ATP utilization of skeletal muscle. One of the questions that arises is how the ATP-producing reactions and pathways are regulated in such a way as to generate ATP strictly according to demand. NMR spectroscopy can contribute in a number of ways to our understanding of these regulatory processes (see Balaban 1990; Radda 1990, 1994). Perhaps the most interesting results relate to the role of ADP in the regulation of oxidative metabolism and tissue function.

It is difficult to measure free ADP directly from ^{31}P spectra, partly because its signals overlap with the much larger signals from ATP, but also because it is apparent that, even if overlap were not a problem, the free ADP is commonly present at such low concentrations that its signals would not be visible above the noise. The evidence for this is obtained by making use of the creatine kinase equilibrium, which permits an indirect measurement of the concentration of free ADP.

Creatine kinase catalyses the following reaction:

$$PCr + ADP + H^+ \rightleftarrows Cr + ATP$$

thereby providing immediate replenishment of the ATP that is broken down during contraction. As described in Section 6.4, ³¹P NMR provides a method of measuring the activity of this enzyme within skeletal muscle non-invasively. In this way, it has been confirmed that in resting muscle the creatine kinase reaction is close to equilibrium, and that the activity of this reaction is sufficiently high to ensure (provided that adequate phosphocreatine is present) that the ATP level is maintained approximately constant during exercise (see, for example, Gadian *et al.* 1981; Rees *et al.* 1988).

The fact that the creatine kinase reaction is close to equilibrium enables the concentration of free cytoplasmic ADP to be calculated, for NMR allows simultaneous measurement of all the other reactants, i.e. phosphocreatine, creatine (from the known concentration of PCr + creatine), ATP, and H⁺. In different types of skeletal muscle at rest, typical values of about 5–20 µM are found for the free ADP concentration. On exercise, analysis of the spectra shows that this concentration can increase to values well above 50 µM. Since it is generally accepted that the K_m for ADP control of mitochondrial ATP synthesis is of the order of 30 µM (see Balaban 1990), ADP provides an obvious candidate for regulating oxidative metabolism in skeletal muscle. However, if the concentration of free ADP rises too much, as it can in disease, it may then begin to inhibit muscle contraction. For example, patients with McArdle's disease fatigue rapidly despite retaining some of their high-energy phosphate, producing very little lactate, and having a normal or slightly alkaline pH (Ross *et al.* 1981). It can be calculated that the concentration of free ADP is unusually high under such circumstances, and this could strongly influence muscle performance (Radda 1990).

A more complete understanding of muscle metabolism and its control *in vivo* needs to take into account any metabolic heterogeneity that may occur at a macroscopic or microscopic level. Developments in spectroscopic imaging now permit the detection of localized variations in muscle metabolites at the macroscopic level (Nelson *et al.* 1991). Plate I shows ³¹P metabolic images of forearm muscle at rest and during exercise, clearly demonstrating heterogeneity in the response to exercise. In comparison with the more conventional surface coil studies, there is necessarily some sacrifice of the temporal resolution that can be achieved; nevertheless, these impressive metabolite maps suggest that spectroscopic imaging will play a complementary role to simple surface coil localization in the investigation of skeletal muscle metabolism and its control.

3.1.4 Energy metabolism and its control in other tissues

³¹P NMR is ideally suited to investigating the large changes in phosphorus-containing metabolites and intracellular pH that are associated with muscular exercise. The large and rapid changes in energy demand and in the concentrations of these metabolites make it reasonable to expect that metabolites such

as ADP should play a key role in the control of energy metabolism, as the above studies have indicated. However, it is less obvious that this should be the case for other tissues such as cardiac muscle, liver. and brain which do not have such extensive variations in energy demand. Indeed, ^{31}P NMR studies of the heart, brain, and kidney have shown little correlation between the rate of oxidative phosphorylation in the steady state and the cytoplasmic concentrations of ADP, P_i, and ATP (Balaban 1990). For example, studies of the normal heart suggest that the phosphorylation potential and the concentrations of ADP and P_i can remain very stable despite large (approximately four-fold) increases in the rates of utilization of ATP (Balaban et al. 1986). This does not mean that these metabolites no longer have the capacity to control oxidative phosphorylation in these tissues. However, it does suggest that, at least under the conditions of these particular studies, there are other factors that presumably have a stronger influence on the regulation of oxidative phosphorylation.

As discussed above, the determination of free ADP concentrations has relied on the use of the creatine kinase reaction. While this method is feasible for skeletal muscle, cardiac muscle, and brain, it cannot be used for tissues that do not express significant amounts of creatine kinase, such as liver and kidney. An interesting means of circumventing this problem has been described by Brosnan et al. (1990), using the transgenic mouse technique to express high levels of creatine kinase in the liver. They were able to show that the free ADP in the liver was about 60 µmol per g wet wt. This is similar to the value previously determined by Veech et al. (1979) on the basis of similar arguments involving enzymes that are believed to catalyse reactions that are near to equilibrium. Such models could clearly contribute significantly to our understanding of tissue metabolism and its control.

^{31}P NMR can make further contributions to our understanding of the kinetics and thermodynamics of oxidative phosphorylation, through its ability to monitor the unidirectional flux between ATP and P_i in intact tissues. As mentioned in Section 2.9, this measurement can be achieved by magnetization transfer techniques, in an analogous manner to the measurement of creatine kinase activity. For example, studies of cardiac muscle (Matthews et al. 1981; Kingsley-Hickman et al. 1987) show that the cytochrome chain is operating under non-equilibrium conditions, which is of direct relevance to mechanisms whereby ATP and, in particular, its products of hydrolysis might exert their control (Balaban 1990).

It should be emphasized that in diseased tissue, there may be aspects of energy metabolism and its control that differ from those seen in normal tissue. For example, in patients with coronary artery disease, ^{31}P spectroscopy showed a reduced PCr/ATP ratio in the anterior left ventricular wall during hand-grip exercise, which was not seen in normal subjects (Weiss et al. 1990). It was suggested that exercise testing with ^{31}P NMR might provide a useful method of assessing the effects of ischaemia on cardiac energy metabolism and of monitoring the response to therapy. Of course, in severe ischaemia, when energy supply is inadequate to maintain the normal tissue energy status, there will be a decline in high-energy phosphates, as discussed in more detail in Sections 3.1.5 and 3.5.

3.1.5 Ischaemia and hypoxia

³¹P MRS provides an excellent experimental approach to the investigation of changes in energy metabolism associated with impaired oxygen delivery, as in ischaemia or hypoxia. Metabolic changes can be monitored sequentially in a single preparation, for example throughout a period of ischaemia and subsequent reperfusion, offering considerable advances over invasive methods of analysis.

One of the main areas of interest is the evaluation of agents or procedures that might protect against the damaging consequences of inadequate oxygen supply. This is of relevance to conditions such as stroke, methods used for the preservation of organs prior to transplantation, and cardioplegic techniques for use during open-heart surgery. Applications to stroke are discussed later in this chapter; here, we briefly discuss problems relating to transplantation and cardioplegia.

Liver transplantation has now become the method of choice for the treatment of end-stage liver disease. A successful outcome relies on a variety of factors, including expertise in organ preservation. The current method of liver preservation is based upon the use of a chilled solution, containing a mixture of sugars, impermeant ions, and pharmacological agents (developed at the University of Wisconsin as UW solution), to flush out blood from the liver, reduce metabolism by cooling, and protect against ischaemic damage (Kalayoglu *et al.* 1988). This allows about 15 h of safe preservation time. However, protection against ischaemic injury is incomplete, since some livers never regain acceptable function after transplantation, whilst in others initial graft function is poor. As a result, there have been many studies attempting to both define and minimize the changes taking place in livers during cold ischaemic preservation. These changes include depletion of high-energy phosphates and acidosis, which can, of course, be assessed non-invasively and continuously using ³¹P spectroscopy (see, for example, Busza *et al.* 1992). Intuitively, one might expect that any procedure that can reduce or slow down the depletion of high-energy phosphates would improve viability, all other things being equal. However, it is clear that the presence or absence of high-energy phosphates is not in itself a sufficient guide to viability; for example in the liver the depletion of high-energy phosphates occurs too early to correlate with viability as assessed by transplantation. In contrast, recovery of ATP following storage may be more informative, being more immediately relevant to the requirements of successful transplantation. This is also amenable to study by ³¹P MRS, both in model systems and *in vivo*, and this may, in due course, lead to a viability test (Bowers *et al.* 1992).

One model system that is particularly suitable to investigation by this type of approach is the isolated heart. For example, Bernard *et al.* (1985) investigated the effects of different cardioplegic solutions on the metabolic state of isolated rat hearts subjected to 2 h of cardioplegic arrest at 15 °C followed by 30 min of reperfusion. Their findings were also correlated with recovery of cardiac function. They were able to conclude that the greatest extent of preservation was provided by a glutamate-containing solution at pH 7.0, and this is one area of research where basic results obtained on animal models have already been transferred to the clinical setting (Bernard and Cozzone 1992).

3.2 ^{1}H MRS OF BRAIN METABOLISM

The observations described in Section 3.1.4 suggest that, while ^{31}P spectroscopy might provide an ideal approach to the investigation of energetics in skeletal muscle, its impact in terms of monitoring changes in the energetics of other tissues might be somewhat smaller. An additional NMR approach involves the use of ^{1}H MRS, which looks particularly attractive for studies of brain metabolism.

The development of ^{1}H NMR for metabolic studies lagged behind ^{31}P NMR for both technical and biochemical reasons. Technically, ^{1}H NMR is more complex than ^{31}P NMR because of the need to suppress the large signals from water and, in some cases, from fats, and because of the large number of metabolites that produce signals in a relatively narrow chemical shift range. However, techniques for solvent suppression and spectral 'editing' are now sufficiently well developed to permit the non-invasive monitoring of many metabolites of interest, provided that the field homogeneity is sufficiently good. For the brain, excellent field homogeneity can often be obtained, so that even at the relatively low (by spectroscopists' standards) field strengths of 1.5–2.0 T that are most commonly used for clinical spectroscopy, adequate spectral resolution can be achieved for many of the signals. In addition, the normal human brain (as opposed to the scalp and bone marrow) generates little, if any, signal from fats, and this facilitates the detection of the co-resonant signal from lactate. For other tissues, these problems are less easily overcome, which explains, at least in part, why most ^{1}H studies of tissue metabolism have focused on the brain.

From a biochemical viewpoint, ^{1}H NMR lagged behind ^{31}P studies because of the perceived strength of ^{31}P NMR in monitoring energy metabolism, as described above. However, it has become apparent that in a number of disorders ^{1}H spectroscopy might reveal abnormalities under circumstances where ^{31}P spectra may appear normal.

A key feature of ^{1}H spectroscopy is the high sensitivity of the ^{1}H nucleus in comparison with other nuclei. In principle, this means that metabolites could be detected at relatively low concentrations. However, it is not necessarily straightforward to observe metabolites at low concentrations, as their signals may be masked by larger signals from other compounds present at higher concentrations. In practice, therefore, the higher sensitivity is generally exploited by trading signal-to-noise ratio for spatial resolution. The higher sensitivity of ¹H spectroscopy means that adequate signal-to-noise ratios can be obtained from smaller volume elements. For example, the linear spatial resolution for clinical ¹H spectroscopy of the brain is typically 1–2 cm, which is about two-fold superior to the resolution achieved with ³¹P spectroscopy.

In the light of the above comments, it is perhaps not surprising to find that, while the majority of clinical spectroscopy studies of the body and limbs still rely on the ³¹P nucleus, the proton is now the dominant nucleus for brain investigations. We now discuss the type of information that ¹H spectroscopy can provide.

3.2.1 Characteristics of ¹H spectra of the brain

Following earlier studies of metabolism in red blood cells (Brown *et al.* 1977), the first ¹H spectra of intact animals were reported by Behar *et al.* (1983). In their studies of anaesthetized rats, they were able to observe ¹H signals from a number of brain metabolites, including *N*-acetylaspartate, creatine + phosphocreatine, and choline-containing compounds, and they demonstrated an increase in the lactate signal on hypoxia, with a return to normal on subsequent oxygenation (Fig. 3.3). Numerous technical developments followed, to the extent that many centres are routinely obtaining spectra of very high quality from well-defined localized regions of the human brain. In addition, an increasing number of groups are exploiting spectroscopic imaging techniques to map the distribution of metabolites within selected planes of the brain, with voxel sizes of about 1 cc. The range of metabolite signals that have been unequivocally identified by ¹H MRS has widened considerably; one notable example is the detection of γ-aminobutyrate (GABA) both in normal subjects and, at elevated levels, in patients treated with the anticonvulsant vigabatrin, which is a specific inhibitor of GABA transaminase (Rothman *et al.* 1993). However, before discussing specific applications of ¹H MRS, it is necessary to discuss some practical issues relating to the appearance and characterization of the spectra.

As mentioned above, a large number of metabolites produce ¹H signals in a relatively narrow chemical shift range, and several approaches have been developed for overcoming the resulting problem of spectral overlap. One commonly adopted approach is to collect data at long echo times (e.g. $TE = 135$ or 270 ms). This has several consequences. First, as explained in Section 8.11, metabolites with complex spin–spin coupling patterns, such as glutamate, tend to produce greatly attenuated signals, and therefore make little contribution to the spectra. Secondly, metabolites with less complex coupling patterns (for example lactate, which generates a doublet centred at 1.32 ppm), produce signals that may be fully inverted at certain echo times. In fact, the main

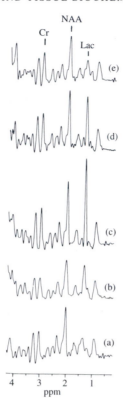

Fig. 3.3 ^{1}H NMR time-course of lactate production and clearance during hypoxia and recovery in the rat brain. The signals from phosphocreatine + creatine (Cr), N-acetylaspartate (NAA), and lactate (Lac) are labelled. The spectra were obtained under the following conditions: (a) normoxic, 25% O_2; (b) hypoxic, 5 min after administration of 4% O_2; (c) after 17 min at 4% O_2; (d) recovery, 14 min after administration of 25% O_2; and (e) after 41 min at 25% O_2. (Reproduced with permission from Behar *et al.* 1983.)

reason for specific selection of echo times of 135 and 270 ms is that a spin-echo sequence with a *TE* value of 135 ms causes the lactate doublet to be inverted, while at 270 ms the signals return to being upright. Singlets (i.e. uncoupled resonances) do not undergo the 'phase modulation' processes that are responsible for these effects, and therefore remain upright, albeit with intensities that are reduced by the effects of T_2 relaxation. A final point is that the spectral components with very short T_2 values (e.g. the relatively broad signals from proteins or lipids) give strongly attenuated signals and therefore tend to make very much smaller contributions as *TE* increases.

These effects are illustrated by the spectra in Figs. 3.4 and 3.5. In all of the spectra, i.e. at all echo times, upright singlet signals are observed from

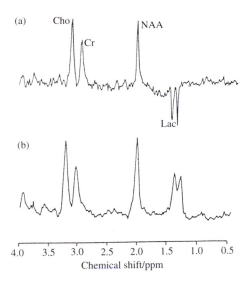

Fig. 3.4 ¹H NMR spectra obtained from two patients with disorders of oxidative metabolism, showing signals from *N*-acetylaspartate (NAA), creatine + phosphocreatine (Cr), choline-containing compounds (Cho), and lactate (Lac). Spectrum (a) was obtained with a *TE* value of 135 ms (reproduced with permission of Kluwer Academic Publishers, Lancaster, from Cross *et al.* (1993), *J. Inher. Metab. Dis.*, **16**, 800–11), and spectrum (b) was obtained with a *TE* value of 270 ms (reproduced with permission from Detre *et al.* (1991), *Ann. Neurol.*, **29**, 218–21).

N-acetylaspartate, creatine + phosphocreatine, and choline-containing compounds. In contrast, the spectra in Fig. 3.4, which were obtained from patients with disorders of oxidative metabolism, show an inverted lactate signal when a *TE* value of 135 ms is used, but an upright lactate when *TE* is increased to 270 ms. If such singlet and doublet signals give adequate information, then the simplicity of the spectra obtained with these relatively long echo times, together with the resulting ease of analysis, makes this type of data acquisition attractive. The main price to be paid by using long *TE* values is that information is lost about other metabolites. Thus, the spectrum obtained in Fig. 3.5(b), obtained at a short *TE* value, is potentially very much richer in information content, with detectable contributions from additional metabolites including glutamate and *myo*inositol. The acquisition of high-quality spectra at very short *TE* values imposes considerable demands upon magnet performance, but with continued improvements in magnet and gradient design, there is increasing use of *TE* values of 30 ms and below for ¹H spectroscopy of the brain. Additional aspects of the pulse sequences commonly used for ¹H spectroscopy are discussed in Chapter 8.

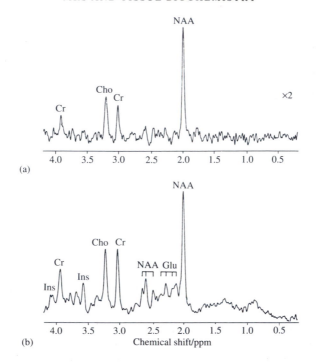

Fig. 3.5 ^{1}H NMR spectra acquired from a normal subject at 2.0 T with *TE* values of (a) 270 ms and (b) 20 ms. The spectra are from a 12 ml volume of white matter, and the signals were assigned as in Fig. 3.4, with additional contributions from glutamate (Glu) and *myo*inositol (Ins). (Reproduced with permission of Raven Press Ltd, New York from Frahm *et al.* (1991*a*), *J. Comput. Assist. Tomogr.*, **15**, 915–22.)

3.2.2 Quantification and assignment of ^{1}H spectra

Detailed interpretation of ^{1}H spectra obviously relies on the ability to assign signals to specific metabolites. In addition, it is necessary to quantify the signals in terms of metabolite ratios, or preferably in terms of absolute concentrations. Some of the methods of spectral assignment were given in Section 2.6, but there are some specific issues relating to ^{1}H MRS of the brain that it is appropriate to discuss here.

Following the approaches described in Section 2.6, many of the assignments in Figs 3.4 and 3.5 are fairly straightforward, provided that we accept certain caveats. The signal at 3.2 ppm (commonly referred to as Cho to reflect choline-containing compounds) contains contributions from a range of N–$(CH_3)_3$-containing compounds, including choline, glycerophosphorylcholine, and phosphorylcholine, as well as betaine. The signal at 3.03 ppm is from creatine +

Human cortex

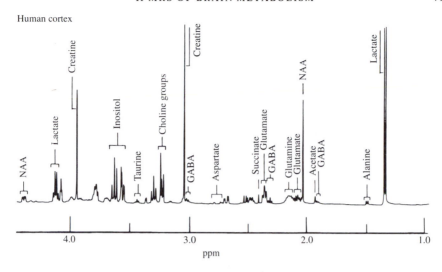

Fig. 3.6 ¹H NMR spectrum obtained at 500 MHz from a perchloric acid extract made from the cortex of human brain tissue obtained at surgery for the amelioration of complex partial seizures. The tissue specimen examined by NMR appeared normal to gross inspection, and light microscopic examination of adjacent samples using routine histological staining revealed no significant pathology. (Reproduced with permission of *Neurology* from Petroff *et al.* (1989), *Neurology*, **39**, 1197–202.)

phosphocreatine, while the signal at 2.01 ppm could, in principle, contain contributions not just from *N*-acetylaspartate but also from other *N*-acetyl-containing compounds. The lactate signal centred at 1.32 ppm overlaps with lipids but can be identified through its characteristic doublet pattern and its associated phase modulation properties (see above). Glutamate and glutamine give rise to complex overlapping multiplets in the region between 2.0 and 2.5 ppm, and the signals in this chemical shift range are often labelled Glx to reflect the combined contributions of these two metabolites. More specific assignment of these signals is possible in some cases, but requires considerable care. The assignment of lipid signals also needs to be treated with some caution, in view of the demonstration that certain brain proteins can give rise to peaks that show some similarities with lipid signals (Behar and Ogino 1993; Kauppinen *et al.* 1993).

Further information about spectral assignments is available from signal intensities, particularly if the intensities can be converted to relative, or preferably absolute, concentrations. For example, there are questions about whether the signal at 2.01 ppm may contain significant contributions from *N*-acetyl-containing compounds other than NAA. These questions have arisen on the basis of measurements of both relative and absolute concentrations (Kreis *et al.* 1993, and refs therein). In particular, the apparent NAA/Cr ratio measured *in vivo* in normal subjects is greater than unity. This contrasts with expectations based on animal data, and also with observations made on human brain extracts (Fig. 3.6), which show an NAA/Cr ratio of less than

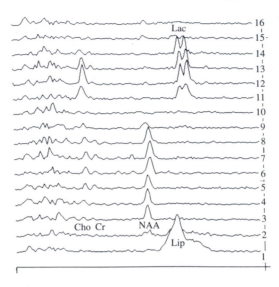

Fig. 3.7 ^{1}H spectra from a patient with a grade II astrocytoma. These are selected spectra from a spectroscopic imaging dataset, showing typical NAA, Cr, and Cho signals from the uninvolved hemisphere (spectra 3–8), and reduced NAA, elevated Cho, and elevated lactate in the tumour (spectra 11–13). The metabolic maps corresponding to this data set are given in Fig. 3.16. (Reproduced with permission of Springer-Verlag GmbH & Co K G from den Hollander *et al.* (1992), *NMR basic principles and progress.* Vol. 27, pp. 151–175.)

unity. N-Acetylaspartylglutamate (NAAG) is one additional metabolite that could contribute to the signal at 2.01 ppm, (Frahm *et al.* 1991b), but its concentration is relatively low, and its presence is unlikely to account for the difference between the extract and *in vivo* data. Further studies, including more extensive work on extracts, are required to resolve the issue; in the meantime, while NAA undoubtedly makes the dominant contribution to the signal at 2.01 ppm, in some publications this signal is labelled NA rather than NAA, reflecting *N*-acetyl-containing compounds more generally.

3.2.3 *N*-Acetylaspartate and neuronal damage

The main interest in NAA centres around its potential use as a 'neuronal marker'. As discussed in Section 2.10, the function of NAA remains uncertain, but there is increasing evidence that, at least in the mature brain, it is present primarily within neurones (Moffett *et al.* 1991; Urenjak *et al.* 1992, 1993 and refs therein), and that a loss of NAA can be interpreted in terms of neuronal loss or damage. In accordance with this, a wide range of ^{1}H MRS studies have shown a reduction in signal intensity at 2.01 ppm that is entirely consistent with the neuronal/axonal loss or damage that might be anticipated on clinical grounds. This signal loss is commonly observed, for example, in stroke (Bruhn *et al.* 1989; Petroff *et al.* 1992), gliomas (Luyten *et al.* 1990; see also Fig. 3.7), and AIDS (Menon *et al.* 1990; Chong *et al.* 1993), all of which are known to be associated with neuronal loss. This certainly appears to be a powerful, non-invasive approach to the assessment of neuronal loss or damage, as further illustrated by the examples outlined briefly in the following sections; nevertheless, a degree of caution is necessary with this interpretation, for several reasons.

First, as mentioned above, it is necessary to consider the possible contributions that metabolites other than NAA may make to the signal at 2.01 ppm. Secondly, the concentration of NAA increases in the developing brain (Bates *et al.* 1989; van der Knaap *et al.* 1990; Toft *et al.* 1994), and this needs to be taken into account in the interpretation of spectra obtained from infants and young children. In relation to this, the presence of NAA in O-2A progenitor cells (see Section 2.10) could also influence interpretation of spectra from this age-group. Thirdly, as with any MRS study, care needs to be taken to ensure that spectral changes reflect changes in concentrations rather than changes in relaxation times. While relaxation phenomena are unlikely to account for selective reductions in the NAA signal intensity, more measurements of T_1 and T_2 in the relevant pathologies would be desirable. Finally, there may be situations in which NAA metabolism changes independently of neuronal loss or damage.

Bearing in mind these reservations, it is reasonable to conclude that, at least in the mature brain, the observation of an abnormally low signal at 2.01 ppm will generally reflect a loss of NAA indicating neuronal (including axonal) loss or damage. Applications in tumours and in stroke are discussed in more detail in Sections 3.4 and 3.5, while two further areas of interest are multiple sclerosis and epilepsy.

3.2.3.1 *NAA and other metabolites in multiple sclerosis*

With recent developments in spectroscopic techniques, ¹H MRS has become more extensively used for the investigation of multiple sclerosis (MS), and can clearly complement MRI in studies of this disorder. The non-invasive detection of MS plaques provided one of the early success stories of clinical MRI (Young *et al.* 1981), and contrast enhancement using Gd-DTPA has proved to be particularly useful for distinguishing between actively inflamed and chronic lesions (Grossman *et al.* 1988; Miller *et al.* 1988). ¹H MRS can provide additional information through a number of signals, including those of NAA, lipids, choline-containing compounds, and *myo*inositol. Several groups (see, for example, Matthews *et al.* 1991*b*) have reported a decrease in the NAA/Cr ratio in patients with MS, which is interpreted in terms of axonal loss or damage. In addition to contributing to our knowledge of the underlying pathophysiology of MS, such MRS measurements may provide a means of assessing lesion load and progression of disease. As such, the technique could complement MRI in the evaluation of therapeutic trials. Another common finding is an elevation in the Cho/Cr ratio which, on the basis of animal studies may, at least in part, reflect an increase in betaine (Brenner *et al.* 1993;

Preece *et al.* 1993). In the studies of Brenner *et al.* (1993), the increased Cho/Cr ratio appeared to be associated with inflammation. There is also considerable interest in the detection of myelin breakdown products through the appearance of signals in the lipid region of the spectrum. While care is needed to ensure that methyl and methylene signals do not reflect contamination from lipids in the scalp, recent reports now provide convincing evidence that lesions generate lipid-like signals (Davie *et al.* 1993; Koopmans et al. 1993). An increased *myoi-nositol* signal has also been reported (Koopmans *et al.* 1993). While the significance of these biochemical changes remains uncertain, they are consistent with abnormal lipid metabolism. It seems reasonable to expect that successful treatment will be accompanied by spectral changes, and that MRS could therefore provide a useful non-invasive means of assessing the response to therapy.

3.2.3.2 *NAA in epilepsy*

Epilepsy is another disorder for which combined MRI/MRS studies look very productive. Epilepsy is a common disorder, and while it is controlled in many cases by medication, some patients continue to have recurrent seizures. If the seizures in such patients arise from a localized region of the brain, there is scope for treatment by surgical removal of the seizure focus. Pre-surgical evaluation requires determination of the seizure focus, an assessment of the underlying pathology, and a knowledge of whether the rest of the brain is normal.

In patients with intractable temporal lobe epilepsy, the most common underlying pathology is hippocampal sclerosis. As a result of recent developments in MRI (including volumetric measurements, optimized use of imaging planes and sequences, and quantitative relaxation time measurements), it is now possible to visualize hippocampal pathology reliably through non-invasive MRI techniques (Jackson *et al.* 1993 and refs therein). ^{1}H MRS can complement these studies by using the NAA signal to detect more diffuse neuronal damage. In patients with intractable temporal lobe epilepsy, it has been shown that a high proportion of spectra obtained from the medial region of the temporal lobe ipsilateral to the seizure focus reveal markedly decreased NAA signals relative to other metabolite signals (Hugg *et al.* 1993; Cendes *et al.* 1994; Connelly *et al.* 1994). A decrease in NAA presumably reflects neuronal loss or damage, and since the spectra are obtained from fairly large regions, the implication is that the damage is not confined to the hippocampus; it extends to a much greater region of the temporal lobe. This is consistent with data from functional imaging techniques (positron emission tomography, PET, and single photon emission computed tomography, SPECT), which also show relatively widespread abnormalities. In fact, the study of Connelly *et al.* (1994) showed that the decreased ratio of NAA relative to the Cho and Cr signals was not simply due to a decline in the NAA signal. There was also an increase in the Cho and Cr signals which, on the basis of studies of purified neural cell types (Urenjak *et al.* 1993), may reflect gliosis.

Some of the patients also showed spectral abnormalities in the contralateral temporal lobe, indicating the presence of bilateral damage. The pathophysiological basis and functional significance of these observations clearly requires further study; meanwhile, it is apparent that ¹H MRS can contribute to the non-invasive assessment of patients being considered for surgical treatment. In particular, if the lateralization indicated by MRS is concordant with that provided by other techniques, it provides additional evidence for the location of the seizure focus, and could help to obviate the need for invasive localization techniques.

3.2.4 ¹H MRS and metabolic disease

As discussed above, one category of ¹H MRS study involves the use of the NAA signal to indicate neuronal loss or damage. Another category involves the detection of more specific spectral abnormalities in metabolic disorders. Interesting observations include the detection of elevated NAA/Cr and NAA/Cho ratios in Canavan's disease, elevated lactate in various disorders of oxidative metabolism, and an increase in glutamine in hepatic encephalopathy and urea cycle disorders. These are discussed briefly below.

3.2.4.1 *Canavan's disease*

Canavan's disease is an autosomal recessive disorder in which spongy degeneration of white matter is observed. Histologically, the affected white matter shows extensive vacuolation and demyelination. Relatively recently, the metabolic basis of this disorder was shown to involve a deficiency in the enzyme *N*-acetylaspartoacylase, which cleaves NAA into acetate and aspartate. Several groups (see, for example, Grodd *et al.* 1990) have shown a large increase in the NAA/Cr and NAA/Cho ratios in children with Canavan's disease, consistent with the enzyme deficiency. However, the mechanisms underlying the impairment in myelination remain uncertain. In particular, if NAA were specifically involved in neuronal metabolism, it would not be at all clear why this enzyme deficiency should result in problems with myelination, which might be expected to be more a reflection of oligodendrocyte dysfunction. The finding that NAA is present in oligodendrocyte precursor cells as well as in neurones (Urenjak *et al.* 1992,1993) provides a possible link between NAA and myelination, and adds to the evidence suggesting that at least one role for NAA is in the generation of myelin.

Further information is available from ¹H MRS analysis of urine specimens from patients with Canavan's disease. ¹H MRS of body fluids has a number of useful features (Iles and Chalmers 1988). In comparison with alternative analytical procedures, no pre-treatment of the sample is required, and, furthermore, no guesses need to be taken as to which metabolites to monitor; ¹H MRS simply monitors those metabolites that are present in sufficiently mobile form and at sufficient concentration to give detectable signals. As

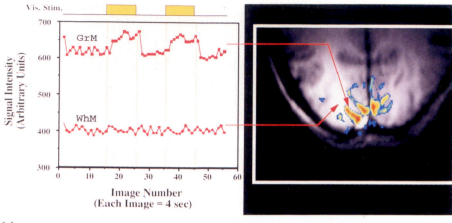

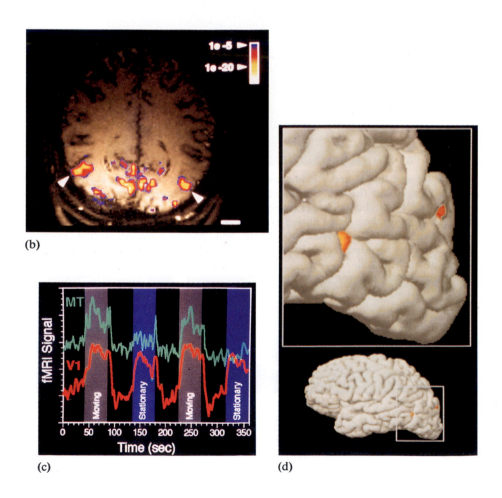

(a)

(b)

(c)

(d)

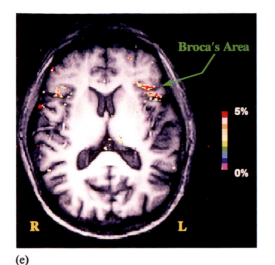

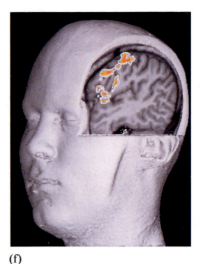

(e)   (f)

Plate III Illustrative functional MRI data. The plate shows a number of ways in which functional and structural MRI data can be presented and superimposed. (a) Functional image obtained at 4 T, showing activated regions of the visual cortex on visual stimulation. Signal intensity changes are shown for grey matter (GrM) and for white matter (WhM). (Courtesy of K. Ugurbil, University of Minnesota.) (b)–(d) Functional images obtained at 1.5 T, obtained with a protocol consisting of alternate presentations of still and moving visual grids, separated by rest periods of darkness. (b) Map of functional data superimposed on a high resolution structural MR oblique axial scan. Areas responding specifically to the moving grid are indicated by arrows. (c) Plots of time course data of regions of interest in the cortical areas V1 and V5 (or MT), showing the lack of response in V5 to a static object. The changes seen in the figure represent increases of about 3% in signal intensity. (d) Areas V5 shown in colour and superimposed on a surface-rendered brain. (Courtesy of B. Rosen and R. Tootell, Massachusetts General Hospital.) (e) Functional MRI data obtained at 4 Tesla, showing activation of Broca's area in a subject performing a silent word generation task. (Courtesy of K. Ugurbil, University of Minnesota.). (f) Another representation of functional MRI data (also obtained at 4 Tesla) from a subject performing a silent-word generation task. The statistically significant functional data are shown in colour overlaid on a grey-scale, volume-rendered three-dimensional MRI of the subject. The activated areas include Broca's area and premotor cortex. (The subject's cheek was flattened by a restraining pad preventing head movement.) (Reprinted with permission of IOP Publishing Limited, from Turner and Jezzard (1994), *Physics World*, August 1994, 29–33.)

(ornithine carbamoyl transferase deficiency) it was found that, even with long echo times ($TE = 135$ ms), glutamine could be detected in the brain during episodes of acute hyperammonaemic encephalopathy with focal neurological abnormalities (Connelly *et al.* 1993). The brain glutamine concentrations were evidently very high, consistent with the hypothesis that intracerebral accumulation of glutamine may contribute to the encephalopathy associated with hyperammonaemia.

3.2.4.4 *Discovery of unknown metabolic disorders*

As mentioned at the beginning of Chapter 2, an important feature of NMR is that it is non-specific, in the sense that signals may be observed from a large number of compounds, without prior specification of which species should be measured. NMR therefore offers the opportunity for detecting metabolic abnormalities that might otherwise go undetected. An excellent example of this is provided by a recent study of a child with an extrapyramidal movement disorder and inconclusive metabolic disturbances, including extremely low creatinine concentrations in serum and urine (Stockler *et al.* 1994). ^{1}H MRS studies of this patient revealed a generalized depletion of creatine in the brain (Fig. 3.9(a)). Oral administration of L-arginine for 4 weeks produced no change in the Cr signal. However, the subsequent administration of creatine (creatine + arginine for 6 weeks, followed by creatine for 6 weeks) led to a significant increase in the brain Cr signal (Fig. 3.9(b)) and a normalization of creatinine in the serum and urine. Similarly, ^{31}P MRS of the brain showed no detectable phosphocreatine before oral administration of creatine, with a significant increase in phosphocreatine afterwards. The partial restoration of creatine levels in the brain was accompanied by an improvement in the patient's condition. On the basis of these and further studies, it was concluded that the patient had an inborn error of creatine synthesis at the level of guanidino-acetate methyltransferase.

This demonstration of a previously unknown metabolic disease provides a particularly striking illustration of the role that MRS can play in elucidating metabolic abnormalities in brain disease, and in relating these abnormalities to neurological deficits.

3.3 ^{13}C MRS AND INTERMEDIARY METABOLISM

As discussed in Section 2.3, the natural abundance of ^{13}C is only 1.1 per cent and this has led to the emergence of two categories of investigation. In one, the compounds of interest are present at sufficiently high concentrations to permit detection without ^{13}C-labelling. Such natural abundance studies are primarily of storage compounds, in particular of glycogen and fatty acids (see Fig. 3.10), since these tend to be present in the appropriate concentration range. The second category exploits the use of ^{13}C-labelling for the investigation

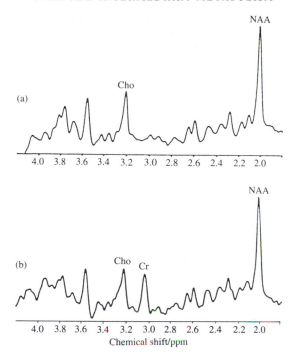

Fig. 3.9 ¹H NMR spectra obtained from a grey matter region of a child with creatine deficiency in the brain (a) prior to treatment, and (b) following oral administration of creatine (creatine + arginine for 6 weeks, followed by creatine for 6 weeks). (Reproduced by permission of *Pediatic Research* from Stockler *et al.* (1994), *Pediatr. Res.*, **36**, 409–13.)

of specific metabolic pathways. We begin by discussing studies involving the detection of glycogen.

In the early 1980s, it was reported that glycogen gives a well-resolved ¹³C NMR spectrum in which the carbons are almost completely NMR-visible (Sillerud and Shulman 1983; see also Gruetter *et al.* 1994, for more recent studies demonstrating the NMR visibility of the liver glycogen signals). This finding was somewhat surprising in view of the fact that large molecules with relatively little mobility tend to give very broad signals. Presumably, glycogen has a high degree of internal mobility such that its carbons give fairly narrow signals. The high concentration of glycogen in liver and muscle means that its ¹³C signals can be detected in these tissues without any need for isotopic enrichment (Jue *et al.* 1989; Rothman *et al.* 1991*b*). For example, it has been shown that ¹³C NMR can distinguish patients with glycogen storage diseases from normal subjects by measuring their muscle glycogen at rest (Jehenson *et al.* 1991).

The use of ¹³C-labelling can provide further information about glycogen metabolism. For example, ¹³C NMR has been used to investigate the disposal

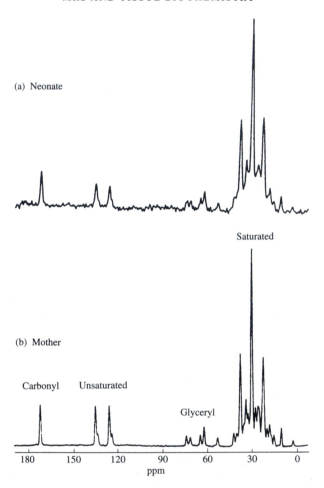

Fig. 3.10 ^{13}C NMR spectra obtained from adipose tissue *in vivo*, showing less unsaturated fatty acid carbons in a neonate (a) than in the mother (b). (From Thomas *et al.* 1994.)

of glucose in normal and non-insulin-dependent diabetic humans (Shulman *et al.* 1990). In this work, hyperglycaemic–hyperinsulinaemic clamp studies with ^{13}C-labelled glucose were carried out in healthy subjects and in subjects with non-insulin-dependent diabetes mellitus. The rate of incorporation of intravenously infused [1-^{13}C]glucose into muscle glycogen was measured in the gastrocnemius muscle. It was concluded that muscle glycogen synthesis provides the main pathway for disposal of glucose in both the normal and diabetic subjects, and that defects in muscle glycogen synthesis play a major role in the insulin resistance that occurs in patients with non-insulin-dependent diabetes mellitus.

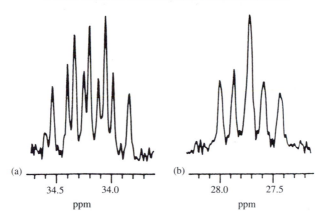

Fig. 3.11 ¹³C NMR spectra from extracts of rat hearts that had been perfused for 30 min with [1,2-¹³C]acetate, [3-¹³C]lactate, and glucose. Only the glutamate C-4 (a) and glutamate C-3 (b) resonances are shown. (From Malloy *et al.* 1990.)

The incorporation of ¹³C-label into glycogen provides a simple illustration of how ¹³C-labelling can be used in the study of intermediary metabolism. The situation rapidly becomes more complex, however, when one considers the spectral characteristics of even relatively simple molecules that have more than one of their carbons labelled with ¹³C (see Jeffrey *et al.* 1991). For example, Fig. 3.11 shows portions of the ¹³C NMR spectra obtained from extracts of rat hearts that had been perfused for 30 minutes with [1,2-¹³C]acetate, [3-¹³C]lactate, and glucose. Only the glutamate C-4 and C-3 signals are shown. In order to understand why such complex signals are obtained, it is necessary to consider what happens as the glutamate becomes multiply labelled.

The proton-decoupled ¹³C spectrum of [4-¹³C]glutamate (i.e. glutamate labelled only at the C-4 position) gives a single resonance. However, if the adjacent C-3 carbon is also ¹³C-labelled, then the signal from C-4 will be split by spin–spin coupling with the C-3 carbon into two signals of equal intensity. Similarly, labelling of the C-5 carbon would also split the C-4 signal into two components, but with a different coupling constant (i.e. frequency separation). Thus [3,4,5-¹³C]glutamate would result in a C-4 signal consisting of four peaks (a doublet of doublets). A mixture of [4-¹³C]-, [3,4-¹³C]-, [4,5-¹³C]- and [3,4,5-¹³C]glutamate could therefore give rise to a total of nine (i.e. 1 + 2 + 2 + 4) signals, as seen in Fig. 3.11(a). The question then arises as to what spectral characteristics can be expected for a tissue presented with, for example, [2-¹³C]acetate. This would enter the citric acid cycle as [2-¹³C]acetyl-CoA (see Fig. 3.12), but would generate oxaloacetate labelled at either C-2 or C-3 as a result of scrambling of the label in succinate and/or fumarate. This oxaloacetate could then condense with more labelled acetyl-CoA, the subsequent labelling patterns becoming quite complex. In practice,

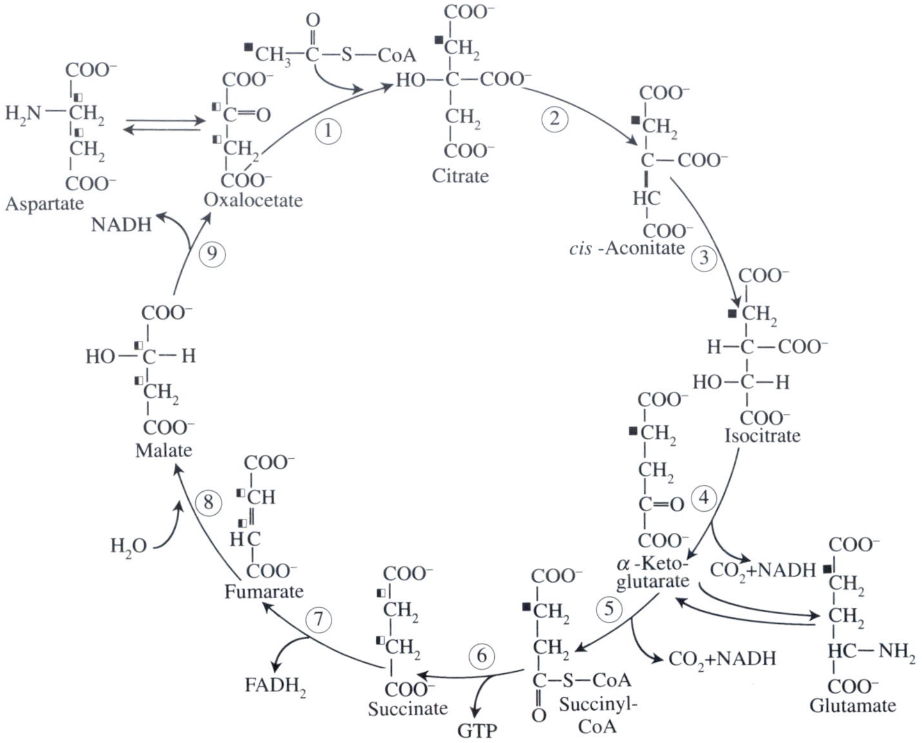
Aspartate
Oxalocetate
Citrate
cis-Aconitate
Isocitrate
α-Keto-glutarate
Glutamate
Succinyl-CoA
Succinate
Fumarate
Malate
NADH
H_2O
$FADH_2$
GTP
CO_2+NADH
CO_2+NADH
1
2
3
4
5
6
7
8
9

Fig. 3.12 Flow diagram showing ^{13}C-isotope distribution during the first turn of the citric acid cycle with [1,6-^{13}C]glucose, [3-^{13}C]lactate, or [2-^{13}C] acetate as the supplied substrate. The filled squares indicate ^{13}C-labelled carbon, the half-filled squares indicate half the level of enrichment. During subsequent turns of the cycle, the glutamate becomes multiply labelled, as a result of scrambling of label in succinate and/or fumarate. (Adapted from Jeffrey *et al.* (1991), *Trends Biochem. Sci.*, **16**, 5–10, with permission of Elsevier Trends Journals.)

this sequence of events enables glutamate to be labelled at the C-3, C-4, and C-5 positions, producing spectra of the type shown in Fig. 3.11. Computer simulations permit identification of how much of the acetyl-CoA entering the citric acid cycle is derived from acetate, how much from lactate, and how much from glucose (Malloy *et al.* 1990), thus providing a direct measure of which of the substrates are selected for oxidation.

Similar studies have been carried out of cerebral metabolism. For example, Künnecke *et al.* (1993) have investigated the metabolism of [1,2-$^{13}C_2$]glucose and of multiply labelled 3-hydroxybutyrate in rat brain, obtaining spectra both *in vivo* and *in vitro*. Figure 3.13 shows *in vivo* ^{13}C spectra of rat brain before infusion (a), and during infusion of each of the two substrates (b and c), while Fig. 3.14 shows corresponding spectra obtained from perchloric acid extracts

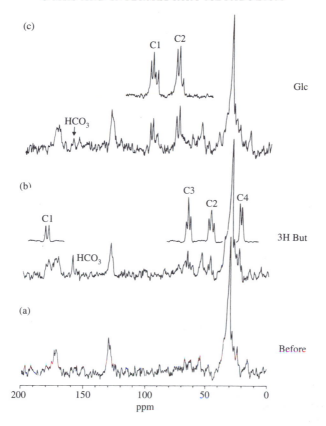

Fig. 3.13 ¹³C NMR spectra of rat brain obtained at 20 MHz *in vivo* before (a) and during the infusion of [U-¹³C₄]3-hydroxybutyrate (3HBut) (b), or [1,2-¹³C₂] glucose (Glc) (c). The insets show the spectra of [U-¹³C₄]3-hydroxybutyrate (b) and [1,2-¹³C₂] glucose (c) in saline solution. (Reproduced by permission of John Wiley and Sons Limited from Künnecke *et al.* (1993), *NMR Biomed*, **6**, 264–77, where spectral assignments are given.)

of the brain. Analysis of the spin-coupling patterns in such spectra provides detailed information about many aspects of cerebral metabolism, including the pathways of glutamate, glutamine, and GABA synthesis, and reveals various features of metabolic compartmentation that can readily be attributed to the differing metabolic characteristics of neurones and glial cells. These studies can also contribute to our understanding of the role of N-acetylaspartate. In particular, it was found that while free aspartate is ¹³C-enriched to a similar extent to glutamate and glutamine, the labelling of the aspartyl moiety of N-acetylaspartate remained at the natural abundance level. In contrast, the acetyl moiety of N-acetylaspartate was significantly enriched, lending further support to the view that this metabolite may act as an acetyl carrier or as a storage form of acetyl-CoA.

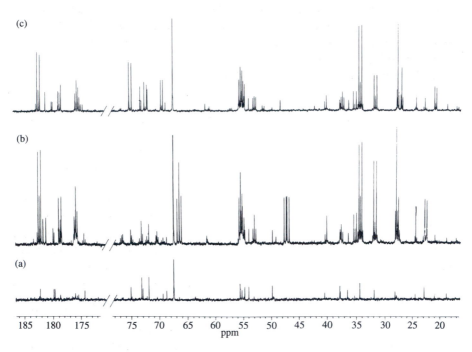

Fig. 3.14 ^{13}C NMR spectra obtained at 100 MHz from perchloric acid extracts of rat brains after the infusion of (a) saline solution, (b) [U-^{13}C$_4$]3-hydroxybutyrate, and (c) [1,2-^{13}C$_2$] glucose. Only the aliphatic and carboxylic regions of the spectra are shown. (Reproduced by permission of John Wiley and Sons Limited from Künnecke *et al.* (1993), *NMR Biomed*, **6**, 264–77, where spectral assignments are given.)

An alternative technical approach exploits the influence of the ^{13}C spin on the signals from neighbouring protons. For example, the CH$_3$ protons of lactate are split into a doublet if the C-3 carbon is ^{13}C-labelled, and the detection of such splitting in ^{1}H spectra has formed the basis for demonstrating turnover of brain lactate in stroke, as discussed in Section 3.5. As an extension of this indirect approach to the detection of ^{13}C-label, the ^{1}H signals can be detected with and without ^{13}C-decoupling, and this provides a means of measuring fractional enrichment of the ^{13}C-label (see Section 8.11). Such 'proton-observe–carbon-edit' methods have been used to measure the rate of label entry from [1-^{13}C]glucose into glutamate. Since the glucose carbon is incorporated into glutamate by rapid exchange with the tricarboxylic acid cycle (TCA) intermediate α-ketoglutarate (see Fig. 3.12), it is possible to use the rate of glutamate-labelling to obtain an estimate of the TCA cycle activity (Mason *et al.* 1992). This method has now been extended from animal studies to humans (Rothman *et al.* 1992; Mason *et al.* 1995), and the time-course of incorporation of ^{13}C-label into C-4 of glutamate was in good agreement with previous PET measurements of oxygen utilization and glucose consumption. These studies, and additional MRS methods for measuring glucose transport and metabolism, have been discussed further in a review by Shulman *et al.* (1993). Since it is intrinsically more sensitive to collect data through the protons rather than by direct detection of ^{13}C signals, it is likely that this type of approach will greatly extend the applicability of ^{13}C-labelling studies *in vivo*.

3.4 MRS AS A PROBE OF TUMOUR METABOLISM

MRS has a number of possible roles in oncology, with applications to a wide range of systems, ranging from body fluids, tissue extracts, and cell cultures to non-invasive studies of tumour metabolism in humans.

One of the more controversial issues arose from the proposal that water-suppressed ^{1}H NMR spectroscopy of plasma might provide an approach to the detection of cancer and the monitoring of therapy (Fossel *et al.* 1986). This proposal, based on the linewidth properties of the resonances from plasma lipoprotein lipids, provoked a great deal of further work and correspondence, the strong consensus of opinion being that there are far too many false positives and false negatives for the linewidth measurement to provide a useful test for cancer (see Shulman 1990). It has also been reported that ^{1}H MRS can distinguish between normal and malignant tissue by the detection of neutral lipids in, or attached to, the membrane protein (Mountford and Tattersall 1987). On the basis of these and subsequent observations (Lean *et al.* 1993), further analysis of this crowded region of the ^{1}H spectrum may well contribute towards our understanding of the biochemical abnormalities associated with malignant disease. However, as MRS is not a sensitive technique, it seems likely that any tests for markers of cancer will rely on more sensitive techniques.

The prime advantage of MRS lies in its ability to probe tumours non-invasively *in vivo*. A number of nuclei have been exploited, and we begin with ^{31}P studies. In view of the enhanced lactate production that is commonly associated with tumour cells, an interesting observation that has emerged from both animal models and human studies is that the intracellular pH measured by MRS tends to be normal or somewhat alkaline (see Negendank 1992 for review). It should be emphasized that the MRS measurement of pH is based on the chemical shift of the inorganic phosphate signal, and is therefore weighted towards those regions containing most inorganic phosphate. It is conceivable, therefore, that tumours may contain regions of low P_i concentration that are not reported upon by MRS. Nevertheless, these observations, when taken together with the low tumour pH values reported by microelectrodes (which primarily reflect the extracellular space; see Vaupel (1992)), suggest that tumours maintain an unusual pH gradient, the intracellular medium being more alkaline than the extracellular space. This contrasts with the pH gradient maintained by normal cells and, as discussed by Vaupel (1992), may reflect the existence within tumour cells of efficient mechanisms for exporting protons into the extracellular space and for importing proton acceptors into the intracellular compartment. Recent ^{31}P MRS studies have demonstrated that tumours do indeed have a reversed pH gradient, and that this could have a number of important metabolic consequences (Stubbs *et al.* 1994). There could also be consequences for those therapies that may depend upon, or exploit, the special pH conditions that prevail within tumours.

In animal studies, considerable emphasis has been placed on using ^{31}P MRS for the evaluation of response to therapy (Daly and Cohen 1989; Steen 1989). There have been two seemingly conflicting patterns of response, one involving an apparent improvement in energy status, with a loss of inorganic phosphate and increase in phosphocreatine, while the other involves a loss of high-energy phosphates and a rapid increase in inorganic phosphate. While the latter can readily be interpreted in terms of rapid cell death, the former pattern was somewhat unexpected. However, it may reflect a complete destruction of some cells (which therefore generate no signal), with a recovery of the energy status of the remainder because of an increase in energy supply relative to energy demand.

These animal studies opened the way to using ^{31}P MRS in clinical studies to evaluate treatment, the overall rationale being that in addition to providing a non-invasive means of monitoring response to therapy, the biochemical changes detected by MRS might precede other signs of response and therefore provide an early guide to the efficacy of treatment. Comparisons of the various clinical studies that have been reported are not straightforward, primarily because the technology has been continually developing; in particular, the localization techniques used in the early studies were not as highly developed as in more recent investigations, and therefore early studies may be more susceptible to contaminating signals from adjacent tissue. However, a number of interesting points are emerging. In particular, the large changes in energy status seen in animal tumours tend not to be seen in human studies. Instead, the focus of attention has shifted towards the phosphomonoester (PME) and phosphodiester (PDE) signals, which make the other main contributions to the ^{31}P spectra.

In tumours there is now good evidence that the major contributions to the PME signal are from phosphorylcholine (PC) and phosphorylethanolamine (PE) (Daly and Cohen 1989; Steen 1989; Cox *et al.* 1992). Similarly, the phosphodiester region includes contributions from glycerophosphorylcholine (GPC) and glycerophosphorylethanolamine (GPE), but interpretation is not straightforward as mobile lipid components can also generate a large signal in this region of the spectrum (Lowry *et al.* 1992). A common characteristic of tumours is an elevation of the PME signal compared with normal tissue, and since PC and PE are intermediates in the pathway of membrane synthesis, this elevation can be interpreted in terms of alterations in membrane metabolism. However, as discussed by Ruiz-Cabello and Cohen (1992), the full biochemical significance of these observations remains to be established. Regardless of the detailed biochemistry underlying these changes, however, the empirical observation has been made that in human tumours the response to therapy often involves a change in the PME signal; a review of the literature has revealed that a decrease in PME occurred in 38 of 47 cancers that responded to treatment, but in only one of the 14 cancers that did not (Negendank 1992). It is this change that is currently attracting most attention. For example, in serial ^{31}P MRS examinations of a carcinoma of the human breast, the changes in the PME signal that were observed during treatment provided a more sensitive indicator of the response to therapy than volume measurements (Fig. 3.15).

Of the other nuclei that are available for metabolic studies *in vivo*, ^{1}H and ^{19}F are of particular interest. ^{1}H MRS has been used by numerous groups to investigate intracranial tumours, using both single volume and spectroscopic imaging techniques (see, for example, Luyten *et al.* 1990; Alger *et al.* 1990; Bruhn *et al.* 1992). As expected, gliomas generally show a reduced NAA signal, while in some cases there is an elevated lactate signal, indicating abnormal glycolytic metabolism. The Cho/Cr ratio is often elevated. Meningiomas commonly show an elevated alanine/Cr ratio (see, for example, Jungling *et al.* 1993); while the significance of this is unknown, it is presumably a reflection of their cellular composition, as cultured meningeal cells also show a relatively large alanine/Cr ratio (Urenjak *et al.* 1993). One of the questions that arises from such studies is whether the spectral patterns that are observed in different tumour types can be used as a non-invasive aid to tumour classification. While clinical studies are not sufficiently well developed to give any definite indication, data from animals suggest that, at least in principle, this is an area that is well worth pursuing. For example, several different chemometric methods have been used to analyse ^{1}H spectra from perchloric acid extracts of three normal tissues (liver, kidney, and spleen) and five types of rat tumour. On the basis of principal component analysis, cluster analysis, and neural network methods, it proved possible to determine the class of each unknown sample (Howells *et al.* 1992). Of course, the spectra of extracts are far better resolved than spectra obtained *in vivo*, but this type of analysis may also prove useful in improving the discrimination provided by spectra obtained from tumours *in vivo*, and with continued improvements in spectral quality and databases, MRS may in due course provide sufficient information for classification.

One of the problems with spectroscopy of tumours is the presence of metabolic heterogeneity, and for this reason chemical shift imaging methods are particularly appropriate for tumour studies. Such metabolic mapping approaches have indeed shown that brain tumours can display considerable heterogeneity, as indicated by the spectra and maps shown in Fig. 3.16. One possible application of such studies is to help neurosurgeons assess which may be the most appropriate regions for biopsy analysis; it may well be that the regions with the most profound spectral abnormalities should be selected for histological diagnosis.

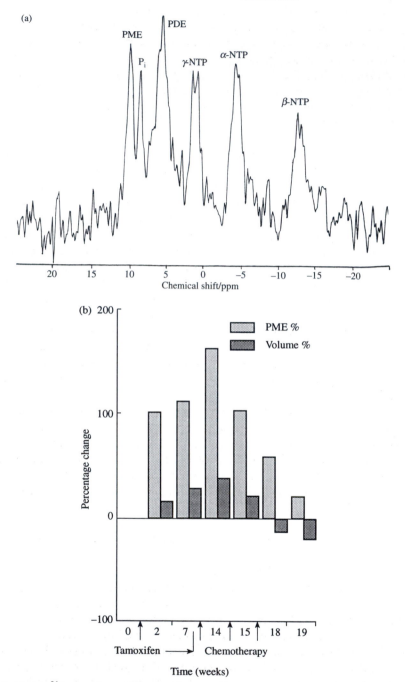

Fig. 3.15 (a) ^{31}P spectrum of a carcinoma of the breast, and (b) diagram showing the PME and tumour volume changes in relation to treatment. (From Glaholm *et al.* 1989.)

^{1}H MRS can also be used for the evaluation of surgical specimens, using similar perchloric acid extraction and analytical procedures to those used in the animal studies described above. Several groups have shown the value of such studies in aiding interpretation of the spectral abnormalities observed *in vivo*. Furthermore, this type of analysis can give additional information about the biochemical characteristics of the different tumour types, and could help to establish the feasibility of using *in vivo* MRS for tumour classification (Gill *et al.* 1990; Peeling and Sutherland 1992; Bruhn *et al.* 1992).

Fluorine-19 is one of the most sensitive of the NMR nuclei, and provides a method for monitoring fluorinated drugs (and anaesthetics) both *in vivo* and *in vitro*. To date, attention has focused primarily on non-invasive studies of the pharmacokinetics and metabolism of 5-fluorouracil *in vivo*. For example, in a recent study of patients who underwent a chemotherapy regimen that included fluorouracil, it was shown that eight out of nine patients who showed intratumoral trapping of the fluorouracil had partial responses to chemotherapy, compared with only two out of 25 patients whose tumours did not trap the drug (Presant *et al.* 1994). It was concluded that ^{19}F MRS can identify patients who are likely to respond to chemotherapy with fluorouracil and that it could be used clinically to help optimize treatment. An important point about ^{19}F MRS is that it permits the simultaneous detection *in vivo* of both the administered drug and its metabolites, providing that they are present at sufficiently high concentrations (about $100\,\mu\mathrm{mol}^{-1}\mathrm{l}$ and above). Such studies can therefore also contribute to our understanding of drug metabolism *in vivo*.

This ^{19}F study provides one example of the various ways in which the metabolic characteristics identified by MRS might provide prognostic indices or predictors of treatment response. These issues have been discussed in a recent review (Negendank 1992), and it is apparent that this is one research area that would benefit greatly from well-controlled multi-centre trials, involving the study of large patient populations over periods of several years.

3.5. MRS IN THE STUDY OF CEREBRAL ISCHAEMIA

Ischaemia refers to reduced blood flow, and has been defined as 'blood flow ... too low to supply enough oxygen to support cellular function' (Lassen and Astrup 1990). MRS has been extensively used to investigate the metabolic changes associated with cerebral ischaemia, anoxia, and hypoxia, both in animal models and in humans. In this final section of the chapter we describe some of the clinical studies that have been undertaken; some animal studies are discussed in Section 4.5 in the context of diffusion-weighted imaging, which provides a powerful new approach to the investigation of pathological processes associated with cerebral ischaemia.

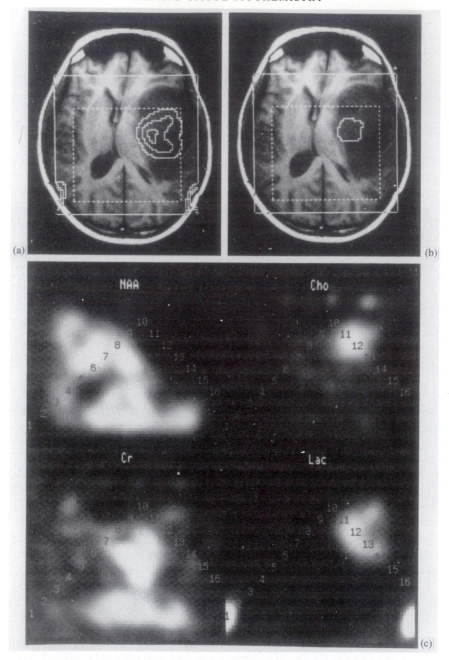

Fig. 3.16 (a) Scout image obtained from a patient with a grade II astrocytoma. The outer box in this image corresponds to the field of view of the spectroscopic imaging data set, and the inner dashed box to the selected volume over which the spectroscopic

Birth asphyxia is the most common cause of impaired neurological function in full-term infants, and hypoxic-ischaemic injury is likely to be the main factor involved. ^{31}P MRS of neonates was started as soon as magnets of large enough bore size became available (Cady *et al.* 1983; Younkin *et al.* 1984), and its sensitivity to abnormalities of oxidative phosphorylation provides scope for investigating the mechanisms underlying hypoxic-ischaemic injury in newborns. In an extensive series of studies, it was shown that the PCr/P_i ratio provides a good prognostic indicator of the likely clinical outcome following birth asphyxia. PCr/P_i ratios within, or close to, the 95 per cent confidence limits for normal controls were shown to be associated with normal outcome or with only minor impairment at 12 months, whereas low PCr/P_i ratios were associated with major neuromotor impairment (Azzopardi *et al.* 1989; Roth *et al.* 1992). The ratio of nucleoside triphosphates to the total NMR-visible phosphate pool was also related to outcome; values below the confidence limits for normal controls were almost always associated with fatal outcome. An important feature of sequential ^{31}P spectra obtained from severely affected infants is that an apparently normal spectrum (in terms of metabolite ratios) is commonly recorded soon after the birth asphyxia episode (8–17 hours), but over the next few days the energy status declines (see Fig. 3.17). This raises the possibility that intervention during the intermediate 'normal' stage may be able to ameliorate the clinical outcome.

While the assessment of metabolite ratios, as in the above study, generally provides a sufficient basis for interpretation, this is not always the case. For example, the normality of metabolite ratios does not always imply that the high-energy phosphate concentrations are normal throughout the area under investigation; it is possible that severely infarcted areas may produce no signal at all. Thus, decreases of up to 40 per cent in the total ^{31}P MRS metabolite signals have been seen in chronic adult infarctions, with no accompanying changes in metabolite ratios or intracellular pH (Bottomley *et al.* 1986). This would be consistent with a substantial loss of viable brain cells, and serves to emphasize the increasing importance that is being placed on measuring absolute, as well as relative, concentrations.

An extensive series of ^{31}P studies have been carried out on patients with acute ischaemic stroke due to major cerebral vessel occlusion (Welch *et al.* 1988). All patients had focal neurological deficits, and they were serially studied at times ranging from 18 hours to 10–40 days after the onset of clinical deficit. Overall, there were distinct metabolic changes that were greatest

---

imaging was performed. The contour lines are a projection from the lactate image of Fig. 3.16(c). (b) The same scout image, but in this case the contours from the Cho signal are indicated. (c) Metabolic maps constructed from the chemical shift imaging data set. Selected spectra from this data set were shown in Fig. 3.7 (Reproduced with permission of Springer-Verlag GmbH & Co. K G from den Hollander *et al.* (1992), *NMR basic principles and progress*, **27**, pp. 151–75.)

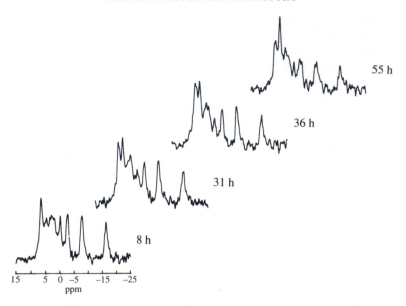

Fig. 3.17 A series of ^{31}P spectra obtained from the cerebral cortex of a birth-asphyxiated infant born at 37 weeks gestation. The post-natal ages at the time of study are indicated. The spectrum obtained at 8 h is similar to that obtained from normal infants. However, the subsequent spectra show a deterioration of energy status. (Reproduced by permission of Pediatric Research from Azzopardi *et al.* (1989), *Pediatr. Res.*, **25**, 445–51.)

during the acute or sub-acute (32–72 hours) stages. No significant abnormalities in high-energy phosphates were observed beyond 7–9 days, despite persistent neurological deficit and imaging evidence for infarction. It was suggested that, apart from partial recovery of some neurones, the return of high-energy phosphates may also originate from glial cells or from macrophages that infiltrate the infarct. pH was initially acidotic, but there were subsequent alkalotic shifts which, it was suggested, may signify the end of active ischaemic cell metabolism and hence define the limit of a therapeutic window.

   These ^{31}P studies of neonates and adults suggest that ^{31}P MRS is of most value as a means of assessing acute damage, but that interpretation of later events is more complex, in part because gliosis and macrophage infiltration might lead to the reappearance of high-energy phosphates in the spectra. It is clearly important to have a means of distinguishing the metabolic properties of different cell types, and in this respect ^{1}H MRS offers greater prospects. In particular, if, as discussed above, N-acetylaspartate is present in neurones but not in mature glial cells, then neuronal loss will result in a loss of the ^{1}H signal from NAA, and this signal would not recover with gliosis or macrophage infiltration. The initial ^{1}H MRS studies of stroke patients clearly showed a loss of NAA (Bruhn *et al.* 1989), and a range of subsequent studies have verified that infarction is indeed associated with loss of NAA, entirely consistent with expectations.

Of additional interest is the detection of lactate in ^{1}H spectra of stroke patients. At its normal concentration of about 0.6 mmol per kg wet wt (Hanstock *et al.* 1988), lactate is barely, if at all, detectable in most ^{1}H spectra of the brain. However, several groups have shown the presence of large lactate signals associated with a variety of pathologies, including stroke (see, for example, Houkin *et al.* 1993; Graham *et al.* 1993). One of the interesting observations has been the persistence of an elevated lactate signal for days or even months following stroke. In order to interpret this persistent elevation of lactate, it is important to determine its cellular localization, and also to establish whether it reflects continued lactate production or a pool of 'trapped' lactate. In one patient studied a month after a cortical stroke, the elevated lactate pool underwent complete metabolic turnover within 60 minutes, as indicated by the rate of incorporation of ^{13}C-label from infused [1-^{13}C]glucose into lactate (Rothman *et al.* 1991*a*). It would be surprising if this continued metabolism were associated with damaged brain cells so long after the onset of stroke. Indeed, a further study suggests that brain macrophages, which begin to appear three days after infarction and gradually disappear over several months, could be a major source of this persistently high lactate (Petroff *et al.* 1992). Taken together, these findings suggest that, while the *early* presence of lactate is presumably a reflection of abnormal brain cell metabolism, interpretation of its *continued* presence needs to take into account the changing cell populations associated with brain damage. More generally, these findings illustrate the point that interpretation of any spectral changes that are observed in diseased tissue requires an appreciation of the cell types that may contribute to such changes.

These various spectroscopic observations complement the many ways in which new MRI techniques can now be used for the investigation of cerebral ischaemia and, as discussed in Section 4.5, the combination of imaging and spectroscopy could prove to be particularly informative in the evaluation of stroke.

REFERENCES

Adams, G. R., Fisher, M. J., and Meyer, R. A. (1991). Hypercapnic acidosis and increased $H_2PO_4^-$ concentration do not decrease force in cat skeletal muscle. *Am. J. Physiol.*, **260**, C805–12.

Alger, J. R., Frank, J. A., Bizzi, A., Fulham, M. J., DeSouza, B. X., Duhaney, M. O., *et al.* (1990). Metabolism of human gliomas: assessment with H-1 MR spectroscopy and F-18 fluorodeoxyglucose. *Radiology*, **177**, 633–41.

Argov, A. and Bank, W. J. (1991). Phosphorus magnetic resonance spectroscopy (^{31}P MRS) in neuromuscular disorders. *Ann. Neurol.*, **30**, 90–7.

Azzopardi, D., Wyatt, J. S., Cady, E. B., Delpy, D. T., Baudin, J., Stewart, A. L., *et al.* (1989). Prognosis of newborn infants with hypoxic-ischaemic brain injury assessed by phosphorus magnetic resonance spectroscopy. *Pediatr. Res.*, **25**, 445–51.

Balaban, R. S. (1990). Regulation of oxidative phosphorylation in the mammalian cell. *Am. J. Physiol.*, **258**, C377–89.

Balaban, R. S., Kantor, H. L., Katz, L. A., and Briggs, R. W. (1986). Relation between work and phosphate metabolites in the in vivo paced mammalian heart. *Science*, **232**, 1121–3.

Bates, T. E., Williams, S. R., Gadian, D. G., Bell, J. D., Small, R. K., and Iles, R. A. (1989). ^{1}H NMR study of cerebral development in the rat. *NMR Biomed.*, **2**, 225–9.

Behar, K. L. and Ogino, T. (1993). Characterization of macromolecule resonances in the NMR spectrum of rat brain. *Magn. Reson. Med.*, **30**, 38–44.

Behar, K. L., den Hollander, J. A., Stromski, M. E., Ogino, T., Shulman, R. G., Petroff, O. A. C., *et al.* (1983). High-resolution ^{1}H nuclear magnetic resonance study of cerebral hypoxia in vivo. *Proc. Natl. Acad. Sci. USA*, **80**, 4945–8.

Bernard, M. and Cozzone, P. J. (1992). MRS of normal and pathological myocardium. In *Magnetic resonance spectroscopy in biology and medicine* (eds J. D, de Certaines, W. M. M. J. Bovee, and F. Podo), pp. 387–409. Pergamon Press, Oxford.

Bernard, M., Menasche, P., Canioni, P., Fontanarava, E., Grousset, C., Piwnica, A., *et al.* (1985). Influence of the pH of cardioplegic solutions on intracellular pH, high-energy phosphates, and postarrest performance. *J. Thorac. Cardiovasc. Surg.*, **90**, 235–42.

Bottomley, P. A., Drayer, B. P., and Smith, L. S. (1986). Chronic adult cerebral infarction studied by phosphorus NMR spectroscopy. *Radiology*, **160**, 763–6.

Bowers, J., Teramoto, K., Khettry, U., and Clouse, M. (1992). ^{31}P NMR assessment of orthotopic rat liver transplant viability. *Transplantation*, **54**, 604–9.

Brenner, R. E., Munro, P. M. G., Williams, S. C. R., Bell, J. D., Barker, G. J., Hawkins, C. P., *et al.* (1993). The proton NMR spectrum in acute EAE: the significance of the change in the Cho:Cr ratio. *Magn. Reson. Med.*, **29**, 737–45.

Brosnan, M. J., Chen, L., van Dyke, T. A., and Koretsky, A. P. (1990). Free ADP levels in transgenic mouse liver expressing creatine kinase. *J. Biol. Chem.*, **265**, 20849–55.

Brown, F. F., Campbell, I. D., Kuchel, P. W. and Rabenstein, D. C. (1977). Human erythrocyte metabolism studies by ^{1}H spin echo NMR. *FEBS Lett.*, **82**, 12–16.

Bruhn, H., Frahm, J., Gyngell, M. L., Merboldt, K. D., Hänicke, W., and Sauter, R. (1989). Cerebral metabolism in man after acute stroke: new observations using localised proton NMR spectroscopy. *Magn. Reson. Med.*, **9**, 126–31.

Bruhn, H., Michaelis, T., Merboldt, K. D., Hänicke, W., Gyngell, M. L., Hamburger, C. *et al.* (1992). On the interpretation of proton NMR spectra from brain tumours *in vivo* and *in vitro*. *NMR Biomed.*, **5**, 253–8.

Burns, S. P., Chalmers, R. A., West, R. J. and Iles, R. A. (1992). Measurement of human brain aspartate *N*-acetyl transferase flux *in vivo*. *Biochem. Soc. Trans.*, **20**, 107S.

Busza, A. L., Fuller, B. J., Lockett, C. J., and Proctor, E. (1992). Maintenance of liver adenine nucleotides during cold ischaemia. The value of a high-pH, high-pK flush. *Transplantation*, **54**, 562–5.

Cady, E. B. (1992). Determination of absolute concentrations of metabolites from NMR spectra. In *NMR basic principles and progress* (ed. P. Diehl, E. Fluck, H. Günther, R. Kerfeld, and J. Seelig), Vol. 26, pp. 249–81. Springer-Verlag, Berlin.

Cady, E. B., Costello, A. M. deL., Dawson, M. J., Delpy, D. T., Hope, P. L., Reynolds, E. O. R., *et al.* (1983). Non-invasive investigation of cerebral metabolism in newborn infants by phosphorus nuclear magnetic resonance spectroscopy. *Lancet*, **i**, 1059–62.

Cendes, F., Andermann, F., Preul, M. C., and Arnold, D. L. (1994). Lateralization of temporal lobe epilepsy based on regional metabolic abnormalities in proton magnetic resonance spectroscopic images. *Ann. Neurol.*, **35**, 211–16.

Chong, W. K., Sweeney, B., Wilkinson, I. D., Paley, M., Hall-Craggs, M. A., Kendall, B. E., *et al.* (1993). Proton spectroscopy of the brain in HIV infection: correlation with clinical, immunologic, and MR imaging findings. *Radiology*, **188**, 119–24.

Connelly, A., Cross, J. H., Gadian, D. G., Hunter, J. V., Kirkham, F. J., and Leonard, J. V. (1993). Magnetic resonance spectroscopy shows brain glutamine in ornithine carbamoyl transferase deficiency. *Pediatr. Res.*, **33**, 77–81.

Connelly, A., Jackson, G. D., Duncan, J. S., King, M. D., and Gadian, D. G. (1994). Magnetic resonance spectroscopy in temporal lobe epilepsy. *Neurology*, **44**, 1411–17.

Cox, I. J., Bell, J. D., Peden, C. J., Iles, R. A., Foster, C. S., Watanapa, P. *et al.* (1992). *In vivo* and *in vitro* ^{31}P magnetic resonance spectroscopy of focal hepatic malignancies. *NMR Biomed.*, **5**, 114–20.

Cross, J. H., Gadian, D. G., Connelly, A., and Leonard, J. V. (1993). Proton magnetic resonance spectroscopy studies in lactic acidosis and mitochondrial disorders. *J. Inher. Metab. Dis.*, **16**, 800–11.

Cross, J. H., Connelly, A., Gadian, D. G., Kendall, B. E., Brown, G. K., Brown, R. M., *et al.* (1994). Clinical diversity of pyruvate dehydrogenase deficiency. *Ped. Neurol.*, **10**, 276–83.

Daly, P. F. and Cohen, J. S. (1989). Magnetic resonance spectroscopy of tumors and potential *in vivo* clinical applications: a review. *Cancer Res.*, **49**, 770–9.

Davie, C. A., Hawkins, C. P., Barker, G. J., Brennan, A., Tofts, P. S., Miller, D. H., *et al.* (1993). Detection of myelin breakdown products by proton magnetic resonance spectroscopy. *Lancet*, **341**, 630–1.

Dawson, M. J., Gadian, D. G., and Wilkie, D. R. (1977). Contraction and recovery of living muscles studied by ^{31}P nuclear magnetic resonance. *J. Physiol.*, **267**, 703–35.

Detre, J. A., Wang, Z., Bogdan, A. R., Gusnard, D. A., Bay, C. A., Bingham, P. M., *et al.* (1991). Regional variation in brain lactate in Leigh syndrome by localized ^{1}H magnetic resonance spectrocopy. *Ann. Neurol.*, **29**, 218–21.

Edwards, R. H. T., Dawson, M. J., Wilkie, D. R., Gordon, R. E., and Shaw, D. (1982). Clinical use of NMR in the investigation of myopathy. *Lancet*, **i**, 725–31.

Fossel, E. T., Carr, J. M. and McDonagh, J. (1986). Detection of malignant tumors. Water-suppressed proton nuclear magnetic resonance spectroscopy of plasma. *N. Engl. J. Med.*, **315**, 1369–76.

Frahm, J., Bruhn, H., Hänicke, W., Merboldt, K. D., Mursch, K., and Markakis, E. (1991*a*). Localized proton NMR spectroscopy of brain tumors using short-echo time STEAM sequences. *J. Comput. Assist. Tomogr.*, **15**, 915–22.

Frahm, J., Michaelis, T., Merboldt, K. D., Hänicke, W., Gyngell, M. L., and Bruhn, H. (1991*b*). On the N-acetyl methyl resonance in localized ^{1}H NMR spectra of human brain *in vivo*. *NMR Biomed.*, **4**, 201–4.

Freeman, D., Bartlett, S., Radda, G. K., and Ross, B. D. (1983). Energetics of sodium transport in the kidney: saturation transfer ^{31}P-NMR. *Biochim. Biophys. Acta*, **762**, 325–36.

Gadian, D. G., Radda, G. K., Brown, T. R., Chance, E. M., Dawson, M. J., and Wilkie, D. R. (1981). The activity of creatine kinase in frog skeletal muscle studied by saturation transfer nuclear magnetic resonance. *Biochem. J.*, **194**, 215–28.

Gill, S. S., Thomas, D. G. T., van Bruggen, N., Gadian, D. G., Peden, C. J., Bell, J. D., *et al.* (1990). Proton NMR spectroscopy of intracranial tumours: *in vivo* and *in vitro* studies. *J. Comput. Assist. Tomogr.*, **14**, 497–504.

Glaholm, J., Leach, M. O., Collins, D. J., Mansi, J., Sharp, J. C., Madden, A., *et al.* (1989). In-vivo ^{31}P magnetic resonance spectroscopy for monitoring treatment response in breast cancer. *Lancet*, **i**, 1326–7.

Graham, G. D., Blamire, A. M., Rothman, D. L., Brass, L. M., Fayad, P. B., Petroff, O. A. C., *et al.* (1993). Early temporal variation of cerebral metabolites after human stroke. *Stroke*, **24**, 1891–6.

Grodd, W., Krageloh-Mann, I., Petersen, D., Trefz, F. K., and Harzer, K. (1990). *In vivo* assessment of N-acetylaspartate in brain in spongy degeneration (Canavan's disease) by proton spectroscopy. *Lancet*, **336**, 437–8.

Grossman, R. I., Braffman, B. H., Brorson, J. R., Goldberg, H. I., Silberberg, D. H., and Gonzalez-Scarano, F. (1988). Multiple sclerosis: serial study of gadolinium-enhanced MR imaging. *Radiology*, **169**, 117–22.

Gruetter, R., Magnusson, I., Rothman, D. L., Avison, M. J., Shulman, R. G., and Shulman, G. I. (1994). Validation of ^{13}C measurements of liver glycogen *in vivo*. *Magn. Reson. Med.*, **31**, 583–8.

Hanstock, C. C., Rothman, D. L., Prichard, J. W., Jue, T., and Shulman, R. G. (1988). Spatially localized ^{1}H NMR spectra of metabolites in the human brain. *Proc. Natl. Acad. Sci. USA*, **85**, 1634–6.

den Hollander, J.A., Luyten, P.R., and Mariën, A.J.H. (1992). ^{1}H NMR spectroscopy and spectroscopic imaging of the human brain. In *NMR basic principles and progress*, (eds. P. Diehl, E. Fluck, H. Günther, R. Kerfeld, and J. Seelig) vol. 27, pp. 151–175.

Houkin, K., Kamada, K., Kamiyama, H., Iwasaki, Y., Abe, H., and Kashiwaba, T. (1993). Longitudinal changes in proton magnetic resonance spectroscopy in cerebral infarction. *Stroke*, **24**, 1316–21.

Howells, S. L., Maxwell, R. J., Peet, A. C., and Griffiths, J. R. (1992). An investigation of tumor ^{1}H nuclear magnetic resonance spectra by the application of chemometric techniques. *Magn. Reson. Med.*, **28**, 214–36.

Hugg, J. W., Laxer, K. D., Matson, G. B., Maudsley, A. A., and Weiner, M. W. (1993). Neuron loss localizes human focal epilepsy by *in vivo* proton MR spectroscopic imaging. *Ann. Neurol.*, **34**, 788–794.

Iles, R. A. and Chalmers, R. A. (1988). Nuclear magnetic resonance spectroscopy in the study of inborn errors of metabolism. *Clin. Sci.*, **74**, 1–10.

Iles, R. A., Stevens, A. N., Griffiths, J. R., and Morris, P. G. (1985). Phosphorylation status of liver by ^{31}P-n.m.r. spectroscopy, and its implications for metabolic control. *Biochem. J.*, **229**, 141–51.

Jackson, G. D., Connelly, A., Duncan, J. S., Grunewald, R. A., and Gadian, D. G. (1993). Detection of hippocampal pathology in intractable partial epilepsy: increased sensitivity with quantitative magnetic resonance T_2 relaxometry. *Neurology*, **43**, 1793–9.

Jeffrey, F. M. H., Rajagopal, A., Malloy, C. R., and Sherry, A. D. (1991). ^{13}C-NMR: a simple yet comprehensive method for analysis of intermediary metabolism. *Trends Biochem. Sci.*, **16**, 5–10.

Jehenson, P., Duboc, D., Bloch, G., Fardeau, M., and Syrota, A. (1991). Diagnosis of muscular glycogenosis by *in vivo* natural abundance ^{13}C NMR spectroscopy. *Neuromusc. Disorders*, **1**, 99–101.

Jeneson, J. A. L., Rodenburg, A. R., de Groot, M., van Dobbenburgh, J. O., van

Echteld, C. J. A., Bar, P. R., *et al.* (1992). Absence of marked intracellular acidosis in the human quadriceps muscle during strenuous dynamic exercise. *Proc. Soc. Magn. Reson. Med. Berlin Meeting*, Society of Magnetic Resonance in Medicine, Berkeley, CA, p. 2710.

Jue, T., Rothman, D. L., Tavitian, B. A., and Shulman, R. G. (1989). Natural-abundance ^{13}C NMR study of glycogen repletion in human liver and muscle. *Proc. Natl. Acad. Sci. USA*, **86**, 1439–42.

Jungling, F. D., Wakhloo, A. K. and Hennig, J. (1993). *In vivo* proton spectroscopy of meningioma after preoperative embolization. *Magn. Reson. Med.*, **30**, 155–60.

Kalayoglu, M., Sollinger, H., Stratta, R., D'Alessandro, A., Hoffman, R., Pirsch, J., *et al.* (1988). Extended preservation of the liver for clinical transplantation. *Lancet*, **i**, 617–19.

Kauppinen, R. A., Niskanen, T., Hakumaki, J., and Williams, S. R. (1993). Quantitative analysis of ^{1}H NMR dectected proteins in the rat cerebral cortex *in vivo* and *in vitro*. *NMR Biomed.*, **6**, 242–7.

Kingsley-Hickman, P. B., Sako, E. Y., Mohanakrishnan, P., Robitaille, P. M., From, A. H., Foker, J. E., *et al.* (1987). ^{31}P NMR studies of ATP synthesis and hydrolysis kinetics in the intact heart myocardium. *Biochemistry*, **26**, 7501–10.

van der Knaap, M. S., van der Grond, J., van Rijen, P. C., Faber, J. A. J., Valk, J., and Willemse, K. (1990). Age-dependent changes in localized proton and phosphorus MR spectroscopy of the brain. *Radiology*, **176**, 509–15.

Koopmans, R. A., Li, D. K. B., Zhu, G., Allen, P. S., Penn, A., and Paty, D. W. (1993). Magnetic resonance spectroscopy of multiple sclerosis: in-vivo detection of myelin breakdown products. *Lancet*, **341**, 631–2.

Kreis, R., Ross, B. D., Farrow, N. A., and Ackerman, Z. (1992). Metabolic disorders of the brain in chronic hepatic encephalopathy detected with H-1 MR spectroscopy. *Radiology*, **182**, 19–27.

Kreis, R., Ernst, T., and Ross, B. D. (1993). Absolute quantitation of water and metabolites in the human brain. II. Metabolite concentrations. *J. Magn. Reson.*, **102**, 9–19.

Künnecke, B., Cerdan, S., and Seelig, J. (1993). Cerebral metabolism of $[1,2-^{13}C_2]$ glucose and $[U-^{13}C_4]$3-hydroxybutyrate in rat brain as detected by ^{13}C NMR spectroscopy. *NMR Biomed.*, **6**, 264–77.

Lassen, N. A. and Astrup, J. (1990). Cerebral blood flow: normal regulation and ischemic thresholds. In *Protection of the brain from ischemia* (eds P. R. Weinstein and A. I. Faden), p. 7. Williams and Wilkins, Baltimore.

Lean, C. L., Newland, R. C., Ende, D. A., Bokey, E. L., Smith, I. C. P., and Mountford, C. E. (1993). Assessment of human colorectal biopsies by ^{1}H MRS: correlation with histopathology. *NMR Biomed.*, **5**, 525–33.

Lowry, M., Porter, D. A., Twelves, C. J., Heasley, P. E., Smith, M. A. and Richards, M. A. (1992). Visibility of phospholipids in ^{31}P NMR spectra of human breast tumours *in vivo*. *NMR Biomed.*, **5**, 37–42.

Luyten P. R., Marien, A. J. H., Heindel, K., van Gerwen, P. H. J., Herholz, K., den Hollander, J. A., *et al.* (1990). Metabolic imaging of patients with intracranial tumours: H-1 MR spectroscopic imaging and PET. *Radiology*, **176**, 791–9.

Malloy, C. R., Thompson, J. R., Jeffrey, F. M. H., and Sherry, A. D. (1990). Contribution of exogenous substrates to acetyl coenzyme A: measurement by ^{13}C NMR under nonsteady-state conditions. *Biochemistry*, **29**, 6756–61.

Mason, G. F., Rothman, D. L., Behar, K. L., and Shulman, R. G. (1992). NMR determination of the TCA cycle rate and α-ketoglutarate/glutamate exchange rate in rat brain. *J. Cereb. Blood Flow Metab.*, **12**, 434–47.

Mason, G. F., Gruetter, R., Rothman, D. L., Behar, K. L., Shulman, R. G., and

Novotny, E. J. (1995). Simultaneous determination of the rates of the TCA cycle, glucose utilization, α-ketoglutarate/glutamate exchange, and glutamine synthesis in human brain by NMR. *J. Cereb Blood Flow Metab.*, **15**, 12–25.

Matthews, P. M., Bland, J. L., Gadian, D. G., and Radda, G. K. (1981). The steady state rate of ATP synthesis in the perfused rat heart measured by ^{31}P NMR saturation transfer. *Biochim. Biophys. Res. Commun.*, **103**, 1052–9.

Matthews, P.M., Allaire, C., Shoubridge, E.A., Karpati, G., Carpenter, S., and Arnold, D. L. (1991a). *In vivo* muscle magnetic resonance spectroscopy in the clinical investigation of mitochondrial disease. *Neurology*, **41**, 114–20.

Matthews, P. M., Francis, G., Antel, J., and Arnold, D. L. (1991b). Proton magnetic resonance spectroscopy for metabolic characterisation of plaques in multiple sclerosis. *Neurology*, **41**, 1251–6.

Menon, D. K., Baudouin, C. J., Tomlinson, D., and Hoyle, C. (1990). Proton MR spectroscopy and imaging of the brain in AIDS: evidence of neuronal loss in regions that appear normal with imaging. *J. Comput. Assist. Tomogr.*, **14**, 882–5.

Miller, D. H., Rudge, P., Johnson, G., Kendall, B. E., Macmanus, D. G., Moseley, I. F., *et al.* (1988). Serial gadolinium enhanced magnetic resonance imaging in multiple sclerosis. *Brain*, **111**, 927–39.

Moffett, J. R., Namboodiri, M. A., Cangro, C. B., and Neale, J. H. (1991). Immunohistochemical localization of N-acetylaspartate in rat brain. *NeuroReport*, **2**, 131–4.

Mountford, C. E. and Tattersall, M.H. (1987). Proton magnetic resonance spectroscopy and tumour detection. *Cancer Surv.*, **6**, 285–314.

Negendank, W. (1992). Studies of human tumors by MRS: a review. *NMR Biomed.*, **5**, 303–24.

Nelson, S. J., Taylor, J. S., Vigneron, D. B., Murphy-Boesch, J., and Brown, T. R. (1991). Metabolite images of the human arm: changes in spatial and temporal distribution of high energy phosphates during exercise. *NMR Biomed.*, **4**, 268–73.

Peeling, J. and Sutherland, G. (1992). High-resolution ^{1}H NMR spectroscopy studies of extracts of human cerebral neoplasms. *Magn. Reson. Med.*, **24**, 123–36.

Petroff, O. A. C., Spencer, D., Alger, J. R., and Prichard, J. W. (1989). High-field proton magnetic resonance spectroscopy of human cerebrum obtained during surgery for epilepsy. *Neurology*, **39**, 1197–202.

Petroff, O. A. C., Graham, G. D., Blamire, A. M., Al-Rayess, M., Rothman, D. L., Fayad, P. B., *et al.* (1992). Spectroscopic imaging of stroke in humans: histopathological correlates of spectral changes. *Neurology*, **42**, 1349–54.

Preece, N. E., Baker, D., Butter, C., Gadian, D. G., and Urenjak, J. (1993). Experimental allergic encephalomyelitis raises betaine levels in the spinal cord of strain 13 guinea-pigs. *NMR Biomed.*, **6**, 194–200.

Presant, C. A., Wolf, W., Waluch, V., Wiseman, C., Kennedy, P., Blayney, D., *et al.* (1994). Association of intratumoral pharmacokinetics of fluorouracil with clinical response. *Lancet*, **343**, 1184–7.

Radda, G.K. (1990). Some new insights into biology and medicine through NMR spectroscopy. *Phil. Trans. R. Soc. Lond. A.*, **333**, 515–24.

Radda, G.K. (1994). Ions, transport and energetics in normal and diseased skeletal muscle. *MAGMA*, in press.

Radda, G. K., Bore, P. J., Gadian, D. G., Ross, B. D., Styles, P., Taylor, D. J., *et al.* (1982). ^{31}P NMR examination of two patients with NADH-CoQ reductase deficiency. *Nature*, **295**, 608–9.

Rees, D., Smith, M. B., Harley, J., and Radda, G. K. (1988). P-31 NMR saturation transfer studies on creatine phosphokinase in human forearm muscle. *Magn. Reson. Med.*, **9**, 39–52.

Ross, B. D., Radda, G. K., Gadian, D. G., Rocker, G., Esiri, M., and Falconer-Smith, J. (1981). Examination of a case of suspected McArdle's syndrome by ^{31}P nuclear magnetic resonance. *N. Engl. J. Med.*, **304**, 1338–42.

Roth, S. C., Azzopardi, D., Edwards, A. D., Baudin, J., Cady, E. B., Townsend, J., *et al.* (1992). Relation between cerebral oxidative metabolism following birth asphyxia and neurodevelopmental outcome and brain growth at one year. *Dev. Med. Child Neurol.*, **34**, 285–95.

Rothman, D. L., Howseman, A. M., Graham, G. D., Petroff, O. A. C., Lantos, G., Fayad, P. B., *et al.* (1991*a*). Localized proton NMR observation of [3-13C]lactate in stroke after [1-13C]glucose infusion. *Magn. Reson. Med.*, **21**, 302–7.

Rothman, D. L., Magnusson, I., Katz, L. D., Shulman, R. G., and Shulman, G. I. (1991*b*). Quantitation of hepatic glycogenolysis and gluconeogenesis in fasting humans with ^{13}C NMR. *Science*, **254**, 573–6.

Rothman, D. L., Novotny, E. J., Shulman, G. I., Howseman, A. M., Petroff, O. A. C., Mason, G., *et al.* (1992). ^{1}H-[^{13}C] NMR measurements of [4-^{13}C]glutamate turnover in human brain. *Proc. Natl. Acad. Sci. USA*, **89**, 9603–6.

Rothman, D. L., Petroff, O. A. C., Behar, K. L., and Mattson, R. H. (1993). Localized ^{1}H NMR measurements of γ-aminobutyric acid in human brain. *Proc. Natl. Acad. Sci. USA*, **90**, 5662–6.

Ruiz-Cabello, J. and Cohen, J. S. (1992). Phospholipid metabolites as indicators of cancer cell function. *NMR Biomed.*, **5**, 226–33.

Shulman, R. (1990). NMR — Another cancer-test disappointment. *N. Engl. J. Med.*, **322**, 1002–3. See also correspondence in *N. Engl. J. Med.* (1990), **323**, 677–81.

Shulman, G. I., Rothman, D. L., Jue, T., Stein, P., DeFronzo, R. A., and Shulman, R. G. (1990). Quantitation of muscle glycogen synthesis in normal subjects and subjects with non-insulin-dependent diabetes by ^{13}C nuclear magnetic resonance spectroscopy. *N. Eng. J. Med.*, **322**, 223–8.

Shulman, R. G., Blamire, A. M., Rothman, D. L., and McCarthy, G. (1993). Nuclear magnetic resonance imaging and spectroscopy of human brain function. *Proc. Natl. Acad. Sci. USA*, **90**, 3127–33.

Sillerud, L. O. and Shulman, R. G. (1983). Structure and metabolism of mammalian liver glycogen monitored by carbon-13 nuclear magnetic resonance. *Biochem. J.*, **221**, 1087–94.

Steen, R. G. (1989). Response of solid tumours to chemotherapy monitored by *in vivo* ^{31}P nuclear magnetic resonance spectroscopy: a review. *Cancer Res.*, **49**, 4075–85.

Stockler, S., Holzbach, U., Hanefeld, F., Marquardt, I., Helms, G., Requart, M., *et al.* (1994). Creatine deficiency in the brain: a new, treatable inborn error of metabolism. *Pediatr. Res.*, **36**, 409–13.

Stryer, L. (1988). *Biochemistry*, 3rd ed. W. H. Freeman, San Francisco.

Stubbs, M., Rodrigues, L., Howe, F. A., Wang, J., Jeong, K. S., Veech, R. L. *et al.* (1994). Metabolic consequences of a reversed pH gradient in rat tumors. *Cancer Res.*, **54**, 4011–16.

Taylor, D. J., Bore, P. J., Styles, P., Gadian, D. G., and Radda, G. K. (1983). Bioenergetics of intact human muscle: A ^{31}P nuclear magnetic resonance study. *Mol. Biol. Med.*, **1**, 77–94.

Taylor, D. J., Crowe, M., Bore, P. J., Styles, P., Arnold, D. L., and Radda, G. K. (1984). Examination of the energetics of aging skeletal muscle using nuclear magnetic resonance. *Gerontology*, **30**, 2–7.

Taylor D. J., Styles, P., Matthews, P. M., Arnold, D. A., Gadian, D. G., Bore, P., *et al.* (1986). Energetics of human muscle: exercise-induced ATP depletion. *Magn. Reson. Med.*, **3**, 44–54.

Thomas, E. L., Bell, J. D., Bryant, D. J., Simbrunner, J., Hanrahan, D., Azzopardi, D., *et al.* (1994). Characterisation of neonatal adipose tissue by *in vivo* ^{13}C NMR spectroscopy. *Proc. Soc. Magn. Reson. 2nd Meeting, San Francisco*, Society of Magnetic Resonance in Medicine, Berkeley, CA, p. 79.

Toft, P. B., Christiansen, P., Pryds, O., Lou, H. C., and Henriksen, O. (1994). T1, T2, and concentrations of brain metabolites in neonates and adolescents estimated with H-1MR spectroscopy. *J. Magn. Reson. Imaging*, **4**, 1–5.

Urenjak, J., Williams, S. R., Gadian, D.G., and Noble, M. (1992). Specific expression of N-acetyl-aspartate in neurons, oligodendrocyte-type2 astrocyte (O-2A) progenitors and immature oligodendrocytes *in vitro*. *J. Neurochem.*, **59**, 55–61.

Urenjak, J., Williams, S. R., Gadian, D. G., and Noble, M. (1993). Proton nuclear magnetic resonance spectroscopy unambiguously identifies different neural cell types. *J. Neurosci.*, **13**, 981–9.

Vaupel, P. (1992). Physiological properties of malignant tumours. *NMR Biomed.*, **5**, 220–5.

Veech, R. L., Lawson, J. W. R., Cornell, N. W., and Krebs, H. A. (1979). Cytosolic phosphorylation potential. *J. Biol. Chem.*, **254**, 6538–47.

Weiss, R. G., Bottomley, P. A., Hardy, C. J., and Gerstenblith, G. (1990). Regional myocardial metabolism of high-energy phosphates during isometric exercise in patients with coronary artery disease. *N. Engl. J. Med.*, **323**, 1593–1600.

Welch, K.M.A., Gross, B., Licht, J., Levine, S. R., Glasberg, M., Smith, M. B. *et al.* (1988). Magnetic resonance spectroscopy of neurologic diseases. *Curr. Neurol.*, **8**, 295–331.

Young, I. R., Hall, A. S., Pallis, C. A., Legg, N. J., Bydder, G. M., and Steiner, R. E. (1981). Nuclear magnetic resonance imaging of the brain in multiple sclerosis. *Lancet*, **ii**, 1063–6.

Younkin, D. P., Delivora-Papadopoulos, M., Leonard, J. C., Subramanian, V. H., Eleff, S., Leigh, J. S., *et al.* (1984). Unique aspects of human newborn cerebral metabolism evaluated with phosphorus nuclear magnetic resonance spectroscopy. *Ann. Neurol.*, **6**, 581–6.

4

Physiological magnetic resonance imaging

The most widespread use of magnetic resonance is in diagnostic radiology; it is the imaging method of choice for examination of disorders of the central nervous system, and is increasingly used for the investigation of diseases in other organ systems (Stark and Bradley 1992; Edelman and Warach 1993). The technique provides excellent contrast and spatial resolution (Fig. 4.1), particularly for the limbs and head, which are least susceptible to motion artefacts. Developments in instrumentation, pulse sequences, and contrast enhancement are generating continued improvements in the diagnostic capability of MRI, and there is every reason to believe that there will be many further improvements in the coming years.

Together with the diagnostic applications of MRI, there is increasing awareness of the role of MRI in the investigation of tissue physiology and function, and it is this role that we emphasize here. To a large extent, these more research-oriented studies exploit the dependence of MRI signals on haemodynamic effects, including blood flow and the oxygenation state of haemoglobin. In addition, there are further MRI approaches such as diffusion-weighted imaging, which are proving particularly sensitive to pathophysiology, and which promise to add significantly to our understanding of a number of disease states, as well as extending the scope of diagnostic MRI.

In this chapter, we describe the main MRI approaches that can be used for physiological studies, and discuss some specific applications, primarily in the heart and brain.

4.1 CONTRAST IN MRI

The contrast in MRI is generated by a number of properties of the protons that give rise to the signals. These include the proton density and the relaxation times T_1 and T_2. It was appreciated at an early stage in the development of clinical MRI that variations in T_1 and T_2 could generate far greater soft-tissue contrast than variations in proton density. This contrast could generally be utilized in an entirely empirical manner, with little need to understand the mechanisms that are responsible for variations in T_1 and T_2. Indeed, we still have a great deal to learn about these mechanisms, although we know, for example, that as the water mobility increases, T_2 also tends to increase (see Section 6.3).

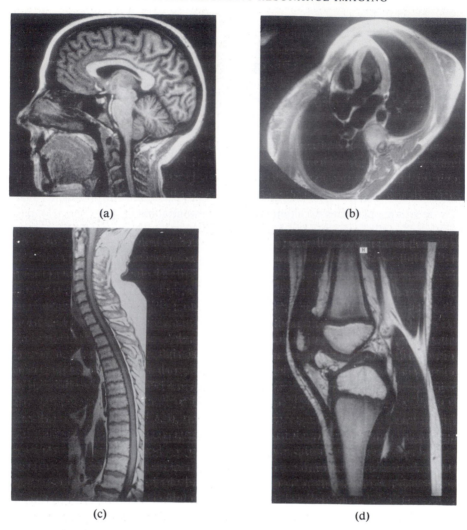

(a)   (b)

(c)   (d)

Fig. 4.1 Images showing the typical anatomical detail that MRI provides. (Courtesy of Great Ormond Street Hospital for Children, Royal Brompton Hospital, and National Hospital for Neurology and Neurosurgery, London.)

For some time, it was hoped that the measurement of T_1 and T_2 values might provide a means of tissue characterization. However, it became apparent that this hope was somewhat optimistic, and that while T_1- and T_2-weighted imaging might provide an excellent means of distinguishing tissues and visualizing pathologies, a greater degree of specificity would be desirable. For example, tumour and surrounding oedema both showed increased signal in T_2-weighted images, but discrimination between the two was sometimes

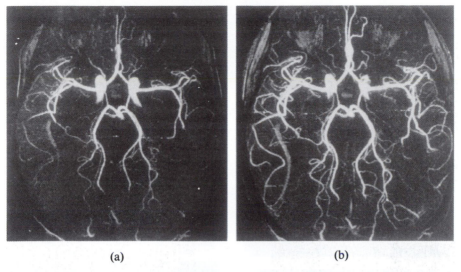

(a)   (b)

Fig. 4.2 Comparison of three-dimensional magnetic resonance angiograms of the human head using the time-of-flight technique without (a) and with (b) the use of magnetization transfer contrast. (Reprinted with permission of Radiological Society of North America from Edelman *et al.* (1992), *Radiology*, **184**, 395–9.)

difficult. As a result, interest was generated in the development of contrast agents, i.e. of pharmaceuticals that could enhance the contrast between different tissues. It had been known since the initial development of magnetic resonance that paramagnetic species can strongly influence the relaxation times of water protons (Bloch 1946), and such species became candidates for increasing the sensitivity, and in particular the specificity, of MRI (for review see Watson *et al.* 1992). Indeed, the gadolinium chelate Gd-DTPA is now extensively used as a contrast agent in diagnostic MRI, and the further development of contrast agents has become a major enterprise, future objectives including improvements in tissue targeting. The contrast generated by Gd-DTPA can also be of value for physiological or functional MRI studies, as described in Sections 4.4 and 4.6.

The sensitivity and specificity of MRI can also be enhanced by taking advantage of additional intrinsic contrast mechanisms. One such approach involves the use of magnetization transfer contrast (Balaban and Ceckler 1992; see Section 6.4), which is contributing to improvements in tissue contrast generally, and to magnetic resonance angiography in particular (Fig. 4.2). Diffusion-weighted imaging provides another interesting approach, particularly in the investigation of acute stroke, as discussed in Section 4.5; for example, the diffusion-weighted images shown in Fig. 4.3(c) and (d) reveal areas of pathology before any significant changes are seen in conventional T_1- and T_2-weighted images (Fig. 4.3(a) and (b)). More generally, it is apparent that, by appropriate

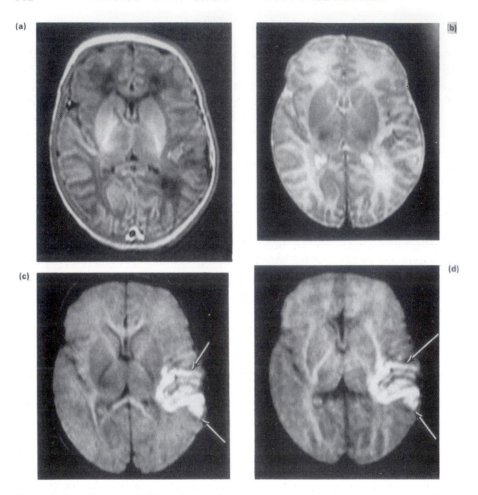

Fig. 4.3 Images of an infant with cerebral infarction. The pulse sequences used were (a) T_1-weighted spin-echo ($TR = 720$ ms, $TE = 20$ ms); (b) T_2-weighted spin-echo ($TR = 3000$ ms, $TE = 120$ ms); (c) and (d) diffusion-weighted, b (see Section 8.5) $= 600$s mm^2. In (c) the diffusion-weighting sensitization is through-plane, whereas in (d) it is left–right. The infarction is difficult to recognize in (a) or (b) but is readily apparent as a high signal on (c) and (d) (arrows). (Reproduced by permission of Chapman and Hall, from Pennock *et al.* 1994).

use of specific pulse sequences, the signal intensities can be influenced by a wide range of specific properties of the protons (or of the molecules to which they belong). Innovative pulse sequences have not only increased the diagnostic capability of MRI, but have also opened up the use of MRI for the study of pathophysiology and function. It is the latter type of study that we concentrate on in the sections below. The pulse sequences that are relevant to these applications are discussed in Chapter 8.

4.2 THE EFFECTS OF MOTION

It was mentioned in Section 2.11 that the sensitivity of the NMR signal to motion can generate artefacts which may degrade image quality; a detailed discussion of these artefacts is included in a review by Henkelman and Bronskill (1987); see also Wood and Ehman (1992). There can be particularly severe problems for studies of the body, because of breathing, peristalsis, and cardiac motion. However, numerous approaches have been developed for suppressing these unwanted effects, and indeed for controlling and manipulating the effects of motion in order to probe physiological processes. In the following sections, we discuss various ways in which haemodynamic and diffusion effects can be exploited, for example, to generate angiograms and to investigate brain function and pathology. Here, we briefly comment on the detection of bulk motion of tissues, in particular of the heart.

High quality images of the heart can be obtained by synchronizing successive data acquisitions with a specific phase of the cardiac cycle (Fig. 4.1(c)). In this way, an image can be built up over, say, 256 cardiac cycles. By collecting data at different phases of the cardiac cycle, impressive movies showing the beating heart can also be generated. However, although this cardiac gating approach strongly attenuates motion artefacts, it does not deal with the adverse effects of respiratory motion, and is not applicable in the presence of arrhythmias. Respiratory gating can be incorporated into the scanning procedure in order to improve image quality, but this prolongs acquisition time and is not used routinely.

Substantial improvements, in particular for functional studies, have resulted from a number of developments in rapid imaging techniques. One approach involves the use of echo-planar imaging (see Stehling *et al*. 1991), which enables a full two-dimensional data set to be acquired following a single radiofrequency excitation pulse; this technique therefore provides a 'snapshot' capability, as data acquisition may take as little as 30–40 ms. An alternative approach involves a number of strategies whereby more conventional imaging sequences can be speeded up. These strategies include the use of low-angle radiofrequency pulses, short repetition times (TR) between successive excitations, and acquisition of data as gradient echoes (Frahm *et al*. 1992b). Such methods can now provide high-resolution cine images of the human heart within a single breath-hold, without any requirement for specialized hardware (Atkinson and Edelman 1991; Sakuma *et al*. 1993; see Fig. 4.4). There are many acronyms for this type of image acquisition; the most commonly used is probably FLASH (fast low-angle shot; Haase *et al*. 1986), and the particularly high-speed variant is often termed TurboFLASH. Numerous variants now exist of both echo-planar imaging and FLASH-type imaging, and there are hybrid techniques that incorporate elements of both. The main points to emphasize are that these techniques help to minimize motion artefacts, and at the same time they can play a key role in the non-invasive visualization and assessment of physiological processes including, for example,

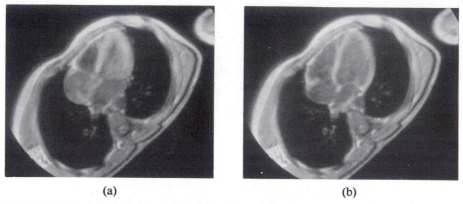

(a)   (b)

Fig. 4.4 (a) End-systolic, and (b) end-diastolic images of a healthy volunteer, obtained with a breath-hold cine sequence. (Courtesy of R. Underwood, Royal Brompton Hospital, London.)

cardiac contractility and blood flow dynamics. Peristaltic activity and fetal imaging are additional areas of potential importance for high-speed imaging.

An alternative approach to imaging heart-wall motion involves spin-tagging. In the technique proposed by Axel and Dougherty (1989), a 'preparation' pulse sequence is incorporated into a standard imaging sequence in order to generate images that show periodic bands or stripes of high and low signal intensity. The intensities of these stripes reflect the magnetization of the water protons that is established by the preparation pulses. The precise location of the stripes is influenced by any movement that occurs between the preparation sequence and the subsequent acquisition of imaging data. In particular, in moving tissue the stripes will move with the tissue, and so this technique can be used for the study of heart-wall motion (see Fig. 4.5), as well as for the investigation of various aspects of blood flow.

4.3 MAGNETIC RESONANCE ANGIOGRAPHY AND FLOW MEASUREMENTS

Magnetic resonance angiography is developing into an increasingly powerful technique. In contrast with conventional angiography, magnetic resonance does not require the injection of an external contrast agent, and furthermore it permits the measurement of flow velocities within vessels. With gradual improvements in techniques and technology, magnetic resonance angiography is becoming increasingly used for the assessment of abnormalities in the cerebrovascular system, and in the abdominal and peripheral vasculature.

Magnetic resonance angiography (MRA) is based on the well-known dependence of signal intensities on bulk flow (Bradley 1992; Masaryk *et al.* 1992).

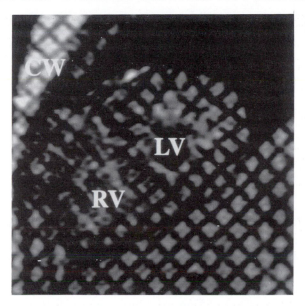

Fig. 4.5 Cardiac image acquired in mid-systole after tagging with a square grid at end-systole. The left ventricle (LV), right ventricle (RV), and chest wall (CW) are labelled. The tags remain square in the chest wall and below the diaphragm, but are deformed in the heart wall due to cardiac contraction. (Courtesy of L. Axel, University of Pennsylvania Medical Center.)

Flow can produce an increase or decrease in signal intensity, depending upon the characteristics of the particular pulse sequence employed. For example, in spin-echo sequences, two radiofrequency pulses are applied, separated by a time $TE/2$, and the signal is detected in the form of an echo that is generated at a time TE after the initial excitation pulse. This echo is produced only by protons that experience both pulses. Blood flow into and out of the selected slice during the period TE can therefore cause loss of signal, because the flowing water protons may experience only one of the two pulses. On the other hand, gradient echo sequences generate signals with a single radiofrequency pulse, and there is therefore no loss of signal resulting from the above mechanism. Instead, if the consecutive excitation pulses are applied at intervals that are much shorter than T_1, there can be an increase in signal intensity because inflowing protons generate more signal than the stationary protons, which undergo partial saturation (i.e. give rise to reduced signal intensity) due to the effects of the rapid pulsing. Such gradient echo methods are now commonly used to generate three-dimensional 'time-of-flight' MR angiograms. These three-dimensional data sets contain information about both stationary tissue and flowing blood, and the vascular structures can be extracted out of the full data set by virtue of their increased signal intensity, in order to generate a display of the vasculature (see Fig. 4.2). This separation of vasculature from surrounding tissue is achieved by a post-processing routine called maximum intensity projection. These projections can be displayed in a cine loop in such a way that the vessels appear to rotate in space. Such cine loops permit visualization of the three-dimensional arrangement of the vessels.

As an alternative to this time-of-flight approach to angiography, there is another type of MR angiography which exploits the fact that flow can induce phase changes in magnetic resonance signals. In particular, as water protons move along a magnetic field gradient, they accumulate a phase shift that is proportional to their velocity (see Section 8.6). Thus, pulsed magnetic field gradients can be incorporated into conventional sequences in order to generate appropriate phase shifts from flowing blood. By varying the magnitude of the gradients, the signals from flowing blood can be altered while those from surrounding tissue remain unchanged. Images displaying the vasculature can therefore be obtained by subtracting images with differing amounts of flow encoding.

In addition to highlighting blood vessels, such phase contrast methods can be used for the assessment of flow rates and velocity distributions within the major vessels. Problems can arise when there is a distribution of flow velocities within a single volume element, as these will cause varying degrees of phase shift with a consequent loss of signal, but there are ways of reducing this loss, or alternatively of exploiting this loss as another means of highlighting the vessels. Compared with time-of-flight techniques, three-dimensional phase contrast methods allow better suppression of background signal, and are more sensitive to slow flow, but involve relatively large scan times. In practice, as is so often the case with magnetic resonance, it is apparent that no single technique is supreme; the choice of technique will depend to a large extent on the question being addressed.

The clinical applications of MR angiography have so far concentrated largely on the head and neck, partly because of the high incidence of cerebrovascular disease, but also because of the relative lack of motion artefact resulting from cardiac and respiratory motion. The MR angiogram shown in Fig. 4.6, which demonstrates a stenosis of the most proximal segment of the basilar artery, illustrates a number of practical features of the technique. On the positive side, the visualization of a vascular abnormality can be achieved non-invasively in a few minutes, within the course of an MRI examination which also provides detailed anatomical information. However, some caution is needed in image interpretation, because of the presence of non-uniform or turbulent flow at or near the region of vessel narrowing. Such flow can result in significant loss of signal and lead to an overestimation of the degree of narrowing (see, for example Wentz *et al.* (1994)). However, techniques for minimizing the effects of turbulent flow are still evolving, and the quality of angiograms continues to improve.

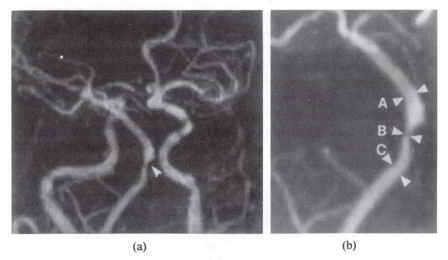

(a)   (b)

Fig. 4.6 Right anterior lateral maximum intensity projection of a patient with occlusion of the right vertebral artery and stenosis of the most proximal segment of the basilar artery (arrowhead on left). The targeted MR angiogram on the right, which gives an improvement in vascular detail, shows a reduction in diameter of about 50% (B compared to A or C), which was greater than that seen on digital subtraction angiography. (Reprinted with permission of Radiological Society of North America from Wentz *et al.* (1994), *Radiology*, **190**, 105–110.)

The intracranial vasculature can also be visualized clearly with MR angiography, as was shown in Fig. 4.2. Applications include the identification of aneurysms, intracranial occlusions, and vascular malformations. Again, however, some caution is necessary; for example, a tight stenosis might, because of the effects of turbulent flow, result in a degree of signal loss similar to that resulting from vascular occlusion. In addition, small vessels with low flow rates may be difficult to visualize. Furthermore, the shortened T_1 values associated with the presence of paramagnetic species, such as Gd-DTPA in a tumour or methaemoglobin in sub-acute haemorrhage, can mimic the effects of flow in time-of-flight images. However, comparison with conventional MRI provides a means of distinguishing these two types of phenomena.

A major achievement for magnetic resonance angiography would be to provide a non-invasive means of assessing the patency of coronary arteries. This has proved to be difficult, primarily because of the artefacts that arise from cardiac and respiratory motion. However, with continued improvements in pulse sequences and machine performance, these difficulties are gradually being overcome, and a recent report demonstrated the visualization of the four major coronary arteries in a series of 39 patients (Manning *et al.* 1993). In this preliminary study, the sensitivity and specificity of MR coronary angiography (compared with conventional angiography) for correctly identifying individual vessels with greater than 50 per cent angiographic stenoses were 90 and 92 per cent, respectively. As an accompanying editorial pointed out (Steinberg 1993), these observations raise the exciting possibility that MR angiography might provide a non-invasive replacement for conventional coronary angiography in some circumstances, but it is still premature to come to any firm conclusions about the eventual clinical scope of this new approach to the assessment of coronary arteries.

Some of the methods that are used for displaying vessels can also provide quantitative assessments of flow velocities. Provided that potential artefacts can be avoided or minimized, the measurements can be accurate and reproducible. Convincing validation of flow measurements has come from cardiac studies in which the left ventricular volume has been measured from multiple contiguous images and compared with the stroke volume as assessed from the aortic flow using velocity mapping (Firmin et al. 1987; Kondo et al. 1991). There was gratifying agreement between the two methods, confirming the reliability of the velocity-mapping measurements. The method has been successfully used for the investigation of abnormal flow in a number of pathological conditions (see Mohiaddin and Longmore 1993 for review).

In principle, the visualization of blood vessels can be enhanced by taking advantage of any magnetic resonance characteristics that distinguish blood from surrounding tissue. Obviously, flow is the key factor, as illustrated by all of the above studies, but if any other factors can also be exploited, then these could provide additional contrast and hence improve the quality of the angiograms. One effect that is proving to be particularly useful is magnetization transfer. As discussed in Section 6.4, the off-resonance application of low-power radiofrequency irradiation can saturate immobile macromolecular protons and, through magnetization exchange processes, reduce the observed signal from water protons. This effect is smaller for blood than for surrounding tissue, and therefore the incorporation of such irradiation into standard angiography pulse sequences can enhance the contrast provided by flow and lead to improved conspicuity of small vessels (Edelman et al. 1992; Pike et al. 1992; see Fig. 4.2).

4.4 MEASUREMENT OF TISSUE PERFUSION

The measurement of tissue perfusion follows logically from the above studies, and is of critical importance to our understanding and assessment of many of the major diseases. There is also considerable interest in the role that perfusion measurements can play in functional neuroimaging (see Section 4.6). From a technical viewpoint, the main problem is that, whereas the above angiography and flow studies involve relatively large vessels and flow velocities, the measurement of tissue perfusion requires an assessment of blood flowing at relatively low velocities through capillaries that are too small to visualize directly. Therefore, somewhat different approaches are necessary.

While a number of different NMR methods for measuring perfusion have

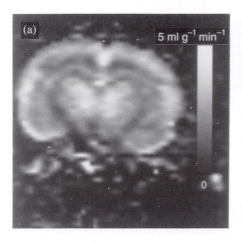
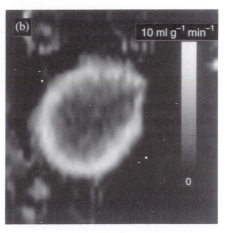

Fig. 4.7 Images showing perfusion of (a) the rat brain and (b) the rat kidney. (Reproduced by permission of John Wiley and Sons Limited, from Detre *et al.* (1994), *NMR Biomed.*, **7**, 75–82.)

been put forward, very few have been able to give any quantitative measures of flow in the established units of blood per 100 g tissue per min. The approach that appears to be most promising involves magnetic labelling of endogenous water protons by saturation or inversion of arterial spins (Zhang *et al.* 1993 and refs therein). For example, if spins in the neck region are selectively inverted, then arterial flow will carry these labelled spins into the brain tissue and hence influence the observed signal from brain water. Comparison of the data obtained with and without this labelling procedure permits measurements of regional perfusion to be made. The signal change resulting from the magnetic labelling depends upon a number of factors, including the perfusion rate, the T_1 value of the tissue water, and any relaxation of the labelled spins during the time they take to travel from the labelling plane to the detection volume.

Figure 4.7(a) shows a calculated perfusion image of a rat brain obtained with this labelling method. The image clearly shows, as anticipated, lower perfusion to the white matter than to the grey matter. Initial animal studies in which changes in cerebral blood flow have been produced by alterations in arterial P_aCO_2 and by bicuculline-induced seizures have indicated that this approach does indeed provide quantitative measurements of regional cerebral blood flow. Further measurements in isolated organs that are perfused at a known rate have confirmed the validity of the method (see Detre *et al.* 1994 and refs therein). Studies have also been carried out on kidney perfusion, the labelling of blood flowing into the kidneys being achieved at either the renal arteries or the descending aorta. As may be seen from Fig. 4.7(b), the calculated image shows high perfusion to the renal cortex with no measurable

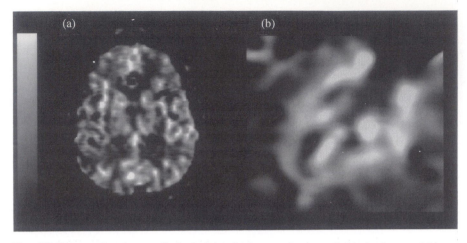

Fig. 4.8 Images showing perfusion of (a) the human brain and (b) the human kidney. The barscale shows cerebral blood flow from $0\text{--}120\,\mathrm{ml\,g^{-1}\,min^{-1}}$ in the case of the brain, and $0\text{--}600\,\mathrm{ml\,g^{-1}\,min^{-1}}$ in the case of the kidney. ((a) Courtesy of Roberts and Detre; (b) Reproduced by permission of John Wiley and Sons Limited, from Detre *et al.* (1994), *NMR Biomed.*, **7**, 75–82.

perfusion to the medulla or papilla; this is because the medulla is supplied largely with blood that leaves the cortex and therefore has magnetization that is in equilibrium with the cortical tissue water. In practice, this means that for renal studies the NMR perfusion imaging method is presently limited to cortical blood flow.

Extension of this arterial spin-labelling method to the investigation of perfusion in humans presents a number of challenges. In particular, the magnitude of the effect produced by inversion of the arterial spins is considerably smaller, for a number of reasons. First, clinical studies are normally performed at field strengths of 1.5 T or below, whereas the animal studies are carried out at higher fields (e.g. 4.7 T). Values of T_1 are lower at the lower field strengths, and this fact alone will reduce the percentage effect that is observed. The percentage effect will be further reduced because the transit times are longer and the regional cerebral blood flow is lower in humans than in rats. Nevertheless, perfusion maps of the human brain and kidney have now been obtained (Roberts *et al.* 1994; Detre *et al.* 1994), as shown in Fig. 4.8. The brain map shows the expected contrast between grey and white matter perfusion rates, the measured perfusion ratio between grey and white matter being about 2.5. Hyperventilation and breath-holding result in appropriate changes in perfusion, and absolute perfusion measurements are consistent with literature values. The renal perfusion images show effects that are analogous to those seen in the rat kidney.

While these results are encouraging, they do require careful attention to a

number of technical and theoretical considerations, as discussed by Detre *et al.*
(1994). These include the effects of patient motion, and the contribution of
effects due to magnetization transfer between water and tissue macromolecules that may be saturated by the labelling pulses. Nevertheless, with continued
improvements in techniques and in the quality of clinical imaging systems, it
seems that this type of approach may become an accepted means of quantify-
ing tissue perfusion non-invasively, one major advantage being that the
method does not require administration of a tracer.

A related method, termed EPISTAR, has recently been used to obtain qualitative maps of cerebral blood flow, with spatial resolution of $2 \times 2 \times 2$ mm
(Edelman *et al.* 1994). This method is an echo-planar imaging modification of
a time-of-flight angiography technique. It involves the alternate acquisition of
two echo-planar images with and without a radiofrequency inversion pulse
applied to inflowing arterial spins. Image subtraction provides a picture of
large proximal vessels when short inflow times are used, while progressively
more distal portions of the tagged vessels are seen as the inflow time is
lengthened. At these longer inflow times, the images represent a qualitative
map of cerebral blood flow. As discussed in Section 4.6, the acquisition of
such images also provides a means of visualizing activated regions of the brain.

An alternative approach to the assessment of perfusion involves the administration of a paramagnetic contrast agent. If a short bolus of contrast agent
is injected intravenously, it is possible to track its passage through the brain
by means of transient changes of signal intensity observed in a series of rapid
images obtained at intervals of, say, 1 (Fig. 4.9). If the arterial supply to any
region of the brain is compromised, this may be detected as a delay or attenuation of the change in signal intensity. A number of studies have been carried
out using Gd-DTPA (Rosen *et al.* 1991; Perman and Gado 1992; Kucharczyk
et al. 1993). Gd-DTPA remains intravascular on passage through the brain,
because it cannot cross the blood–brain barrier. Being paramagnetic, the Gd-
DTPA generates field inhomogeneities through magnetic susceptibility effects.
These inhomogeneities extend beyond the vasculature into the adjacent tissue,
and cause signal intensity loss in T_2^*-weighted images. It is this loss of signal
intensity that provides the basis for the contrast used in these rapid-imaging
studies.

It should be emphasized that, whereas the arterial spin-labelling approach
of Detre *et al.* (1994) is essentially a steady-state method that exploits a freely
diffusible 'tracer' (i.e. magnetically labelled water), the use of paramagnetic
contrast agents involves the detection of transient changes in signal intensity
in response to agents that, in the normal brain, can be regarded as intravascular. As discussed by Weisskoff *et al.* (1993), the quantitative measurement
of cerebral perfusion rates has not yet proved possible using such intravascular
NMR tracers, although in some circumstances a number of haemodynamic
parameters can be assessed, including relative cerebral blood volumes and flows.

Bolus tracking of Gd-DTPA as an indicator of blood flow has even more

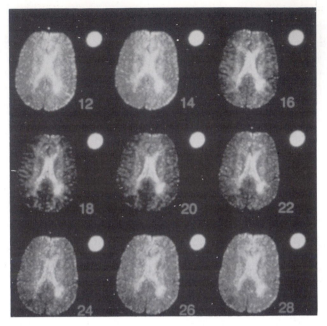

Fig. 4.9 Series of images obtained following administration of Gd-DTPA (times are shown in seconds). The time course of the imaging changes can be used for the qualitative assessment of cerebral haemodynamics. (Courtesy of B. Rosen and R. M. Weisskoff, Massachusetts General Hospital, Boston.)

problems for cardiac studies than for the brain. For the heart, Gd-DTPA is effectively an interstitial agent rather than an intravascular agent (as there is no equivalent of a blood–brain barrier), and so the accumulation of Gd-DTPA in the myocardium depends not only on tissue blood volume and blood flow, but also on the size of the extracellular space and the capillary permeability (Wilke *et al.* 1993). Furthermore, the precise effects of the paramagnetic agent on the signal intensities depend on a number of additional factors, including the exchange rates of water molecules between the various compartments. For these reasons, intravascular agents offer potential advantages, as illustrated by recent investigations of regional myocardial blood volume and flow using the contrast agent polylysine-Gd-DTPA (Wilke *et al.* 1995).

The administration of freely diffusible 'tracers' in the form of deuterium-labelled or [17]O-labelled water, or of exogenous fluorine-containing species, provides another means of assessing perfusion, as discussed in Section 2.11. However, fairly high concentrations need to be used in order to achieve adequate signal-to-noise ratios and/or spatial resolution. Even then, the spatial resolution is necessarily much poorer than that of methods exploiting the ^{1}H signal of water, because the concentrations of the 'tracers' are still much lower than that of water. Furthermore, possible toxicity effects need to be considered.

Finally, considerable interest was raised by the suggestion that it might be possible to model capillary flow in terms of random diffusion-like processes, and that 'diffusion-weighted' imaging might be sufficiently sensitive to distinguish this perfusion effect from the effects of diffusion arising from random Brownian motion (Le Bihan *et al.* 1986). Unfortunately, despite extensive work by several groups, quantitative measurements of tissue perfusion using this method have not proved feasible; the practical and theoretical difficulties are very severe (King *et al.* 1992). However, the interest that was generated in diffusion-weighted imaging was nevertheless of considerable value because, as discussed in the following section, the assessment of water diffusion is proving to be of great interest in the investigation of brain disease.

4.5 DIFFUSION-WEIGHTED IMAGING

As mentioned in Section 2.11, magnetic resonance methods for measuring diffusion were developed by Stejskal and Tanner in the 1960s, and in recent years diffusion measurements have received particular attention as a result of their high sensitivity to pathology. In this section, we discuss the use of diffusion-weighted imaging in the assessment of brain pathology, concentrating on its remarkable sensitivity to early events associated with cerebral ischaemia.

Diffusion results from the thermal translational motion of molecules. It is a random process, often referred to as Brownian motion, and involves displacement distances that are small; for example water molecules, if unrestricted, will typically diffuse in any given direction through a distance of 20 μm in 100 ms, or through 60 μm in 1 s. These distances are comparable to cellular dimensions, raising the possibility that the measurement of water diffusion might provide a means of probing cellular integrity and pathology. In particular, it might be anticipated that if cell structures restrict the diffusion of water molecules in any way, then perturbations in these structures might influence the degree of restriction and hence affect typical diffusion distances.

As described in Section 8.5, images that are sensitized to the diffusional properties of water can be obtained by incorporating pulsed magnetic field gradients into a standard spin-echo sequence. Essentially, these gradients make the images sensitive to the small water displacements that are characteristic of diffusion; in particular, there is a loss of signal because of the random phase shifts acquired by the water protons as a result of their translational motion along the direction of the field gradients. Of course, the images are also more generally sensitive to other types of motion, and so motion artefacts can be especially problematical in diffusion-weighted imaging. This is one reason why diffusion-weighted imaging studies have focused largely on the brain, where there is relatively little artefact due to cardiac and respiratory motion. The other reason is the high sensitivity to certain types of brain pathology.

One of the areas of interest exploits the anisotropy of certain structures. If diffusion is anisotropic, for example if it is restricted in one direction and unrestricted in others, then diffusion-weighted imaging may be able to highlight this anisotropy by comparing signal intensities obtained with the pulsed field gradients applied first in one direction and then in another. This approach has been used for the selective visualization of white-matter tracts, which give less diffusion-weighted signal when the pulsed field gradients are oriented along the direction of the tracts than when the gradients are in the transverse direction (see Fig. 4.3(c) and (d)). These findings indicate that water can move less freely in the direction transverse to the tracts, presumably because the myelin sheaths impede water mobility (see Le Bihan and Turner 1992). The visualization of white-matter tracts offers an additional means of studying functional connectivities in the brain, and of investigating white-matter disorders such as multiple sclerosis.

The aspect of diffusion-weighted imaging that has so far provoked most interest has been its sensitivity to very early events in cerebral ischaemia. The use of MRI to depict these events had been hampered by the apparent lack of early changes in conventional T_1- or T_2-weighted images. However, Moseley *et al.* (1990) showed that diffusion-weighted imaging could be used to detect ischaemic regions of the cat brain before any changes could be detected with conventional MRI protocols. The increases in signal intensity that were seen in diffusion-weighted images indicated that there was a reduction in the apparent diffusion coefficient of water within the ischaemic tissue. They suggested that this may be a reflection of cell swelling, with a shift of water from a relatively unrestricted diffusion environment in the extracellular space to a relatively restricted intracellular environment. It should be emphasized that this *increase* in signal intensity contrasts with much later effects; when necrosis eventually sets in, there is enhanced freedom of movement, with a resulting *decrease* in diffusion-weighted signal intensity.

In subsequent studies of the rat brain, regions of diffusion-weighted hyper-intensity were seen at the time of the first acquired image, 16 min after lesion induction, whereas changes in T_2 were not evident until 2–3 hours later (Mintorovich *et al.* 1991; see also Fig. 4.10). Further studies of cerebral ischaemia have shown that in the gerbil brain diffusion-weighted imaging intensities only increase if the blood flow is reduced to 15–20 ml per 100g per min and below (Busza *et al.* 1992). This is similar to the critical flow threshold for maintenance of tissue high-energy phosphates and ion homeostasis. In addition, with the onset of severe global cerebral ischaemia, diffusion-weighted image intensity increased gradually after a delay of about 2.5 min, consistent with the time course of energy failure and the consequent increase in extracellular K^+. These observations suggest that the diffusion-weighted imaging changes observed early after the onset of ischaemia are sensitive to the disruption of tissue energy metabolism or are a consequence of this disruption, including cell swelling associated with loss of ion homeostasis.

The direct study of ion homeostasis and its relationship to tissue energetics

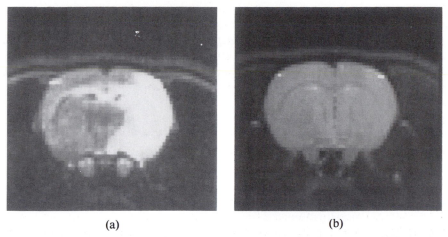

(a)   (b)

Fig. 4.10 Images of the rat brain 1.5 hours following occlusion of the middle cerebral artery. Ischaemic tissue in the territory supplied by the middle cerebral artery is highlighted on the diffusion-weighted image (a) but is not detected in the equivalent T_2-weighted image (b). (Reproduced with permission of Raven Press Ltd, New York, from van Bruggen *et al.* (1994), *Cerebrovasc. Brain Metab. Rev.*, **6**, 180–210.)

can be carried out by combining ^{31}P and ^{23}Na NMR, taking advantage of the fact that the ^{23}Na signals of intracellular and extracellular Na$^+$ ions differ in their relaxation properties (Naritomi *et al.* 1988; Eleff *et al.* 1991; Pekar *et al.* 1991). For example, Pekar *et al.* (1991) showed that upon production of cerebral ischaemia there is a delay of approximately 2 min before the ATP level begins to fall and the intracellular Na$^+$ begins to rise, a finding that correspond very well with the time course of the diffusion-weighted imaging changes described above.

Benveniste *et al.* (1992) have tested the hypothesis that the diffusion-weighted imaging changes are related to the disruption of ion homeostasis and the consequent cell swelling, by administering the glycoside ouabain, which is a specific inhibitor of Na$^+$/K$^+$ATPase. This resulted in diffusion-weighted imaging changes that were very similar to those observed during focal ischaemia. Furthermore, similar effects were observed on addition of the excitoxins glutamate and *N*-methyl-D-aspartate (NMDA), both of which are known mediators of the acute pathology of cerebral ischaemia. It is interesting to note that the apparent diffusion coefficient of brain water has also been observed to fall during status epilepticus (Zhong *et al.* 1993, 1995), a condition in which again there is cell swelling.

While the above findings suggest that the diffusion-weighted imaging changes are related to changes in cell volume, the precise biophysical mechanisms underlying the changes in diffusion-weighted imaging intensity still remain uncertain. Indeed, several alternative explanations have been put forward for the observed signal intensity changes. As discussed by Zhong *et al.* (1993), these include a reduction in membrane permeability and a reduction in energy-dependent cellular motility processes. Although the issue is not completely settled, there is now a strong consensus that the increase in diffusion-weighted imaging intensity is associated with cell swelling (see Kohno *et al.* 1995). This view is based in part upon the observations described above, together with increasingly compelling evidence from additional studies. For example, in a rat model of NMDA-induced brain injury, Verheul *et al.* (1994) have investigated the relationship of changes observed in diffusion-weighted images to electrical impedance measurements. Their results showed that the time course of the intensity changes in the diffusion-weighted images paralleled the progressive shrinkage of the extracellular space as measured by the electrical impedance. After administration of the NMDA antagonist MK-801, the signal enhancement was reversed, in a manner that paralleled the normalization of the impedance measurements.

Further research will no doubt shed light on the biophysical basis for the observed relationships between diffusion-weighted hyperintensities and changes in intracellular and extracellular volumes. Meanwhile, it is apparent that diffusion-weighted imaging is adding significantly to the range of MRI techniques that can be exploited for the non-invasive investigation of tissue pathophysiology. For example, the relationship of the imaging changes to impaired energy metabolism has opened up the possibility of using diffusion-weighted imaging to visualize regions of compromised energy metabolism in humans, with the spatial resolution characteristic of MRI, i.e. with much superior spatial resolution to that provided by ^{31}P or ^{1}H MRS of energy metabolites. Indeed, initial clinical studies are encouraging (Warach *et al.* 1992, 1995; Cowan *et al.* 1994; see Fig. 4.3), and it seems likely that diffusion-weighted imaging will play an important role in improving the early diagnosis of stroke and in the development and evaluation of early stroke interventions. In this respect, the combination of imaging with the spectroscopic techniques discussed in Section 3.5 could be particularly informative.

4.6 FUNCTIONAL NEUROIMAGING

4.6.1 Background

Functional neuroimaging is one of the more remarkable[*] of recent developments in MRI, for it now provides a means of visualizing functionally activated regions of the brain completely non-invasively. It is likely that this new MRI approach, together with complementary neuroimaging methods, will lead to major advances in our understanding of normal brain function and brain dysfunction in disease.

Functional neuroimaging (or brain mapping) refers to the visualization of local physiological changes within the brain that are associated with activation of the visual, motor, or other brain systems. The technique that has so far

contributed most to functional brain mapping is positron emission tomography (PET). These PET studies rely on the detection of changes in local
haemodynamics that are associated with cerebral activation; in particular,
there are changes in regional cerebral blood flow that can be mapped with
PET through the use of ^{15}O-labelled water. The MRI studies of functional
activation that have recently been described also rely on haemodynamic
changes, and we begin by describing the first report of functional brain mapping with MRI (Belliveau *et al.* 1991), which involved the administration of
the paramagnetic contrast agent Gd-DTPA to act as a marker of cerebral
blood volume.

This initial study involved the investigation of the human visual cortex using
a visual stimulus paradigm that, on the basis of PET studies, produces changes
in regional cerebral blood flow of about 30–50 per cent. Echo-planar MR
images were collected at intervals of 750 ms, before, during, and after injection
of Gd-DTPA, which was administered as a bolus into the antecubital vein.
As the Gd-DTPA passed through the brain, it produced a transient reduction
in water signal intensity because of its magnetic susceptibility effects (see
Sections 4.4 and 6.5). From the time course of the signal changes, it was
possible to generate images that reflect relative cerebral blood volume. The
imaging procedure was carried out under control conditions, and also during
visual stimulation produced by goggles that generated patterned flashes at
8 Hz. Subtraction of the control from the activated blood-volume image
revealed those areas in which there was a change in blood volume associated
with the task activation. As shown in Plate II, this corresponds to the primary
visual cortex.

As it happened, this initial activation study was very rapidly superceded by
an alternative MRI approach which had the major advantage that no extrinsic
contrast agent was required. It had been known for some time that there are
several mechanisms whereby changes in blood flow, volume, and oxygenation
can influence signal intensities without any requirement for exogenous agents,
and it soon became apparent that activated regions of the human brain could
be visualized using such intrinsic contrast mechanisms. In these studies, particular interest has focused on T_2^*-weighted signal intensity changes that have
been attributed primarily to changes in local venous blood oxygenation, in
particular to the effects generated by the paramagnetic iron centres of deoxyhaemoglobin. NMR studies of this compound have a long history, and some
of the salient points are described briefly below.

4.6.2 Effects of haemoglobin

On deoxygenation of haemoglobin, the electron spin of the haem Fe^{2+}
changes from a diamagnetic to a paramagnetic state. As discussed in Section
6.5, paramagnetics can influence many properties of NMR signals, through
a variety of mechanisms. For example, if we consider the ^{1}H spectrum of

deoxyhaemoglobin itself, the presence of the paramagnetic Fe^{2+} centres causes significant broadening and shifts of some of the spectral lines of the protein itself. This is of interest in the investigation of the structural properties of deoxyhaemoglobin, and in principle offers an approach to studying tissue oxygenation *in vivo*; indeed, analogous effects in myoglobin have been used to assess oxygenation status in cardiac and skeletal muscle, as mentioned in Section 2.12. Of course, there are many problems with detecting signals from haemoglobin *in vivo*, not least of which is its relatively low overall concentration, and this does not appear to be a promising approach to pursue. What is fortunate is that the proton signals of neighbouring molecules, including those of water, can also be influenced by the paramagnetic Fe^{2+} centres of deoxyhaemoglobin, and as a result it is possible to use the water protons to report on changes in oxygenation status. In particular, there are magnetic suscepibility effects associated with this paramagnetism, as a result of which local field gradients are generated in and around the blood vessels. These cause signal attenuation in T_2^*-weighted images, which is not confined to water within the vessels; it can extend significantly into the surrounding tissue. As a result, there may be substantial signal changes associated with changes in oxygenation state, even if the blood volume is relatively small.

Ogawa and co-workers (1990*a,b*) have explored this effect in a series of animal studies, and have shown that the T_2^*-weighted proton signal intensity changes in the expected manner with alterations in cerebral oxygenation, including alterations caused by pharmacological agents. Similar effects have been described by Turner *et al.* (1991). Thus, if focal changes in oxygenation occur on task activation, one might similarly expect T_2^*-weighted imaging to be sensitive to these changes and therefore provide a means of mapping activated regions of the brain. Evidence that oxygenation changes do indeed occur on activation is available from PET studies, which have shown that during cerebral activity there is an increase in local blood flow with relatively little change in oxygen consumption, so that the venous blood should become more oxygenated on activation (Fox and Raichle 1986; Fox *et al.* 1988). This increase in venous oxygenation, i.e. decrease in venous deoxyhaemoglobin, should cause an increase in T_2^*-weighted signal intensity and, as we shall see, this is in fact observed in activated regions of the brain.

The contrast in T_2^*-weighted functional images is generated because gradient echoes do not refocus the dephasing effects of local field homogeneities associated with the presence of deoxyhaemoglobin. If spin-echoes are used instead of gradient echoes, then such dephasing effects will be refocused, provided that there is no significant diffusion of the water molecules through the field gradients. This explains why T_2^*-weighted images are more sensitive than T_2-weighted images to the effects of brain activation. However, if the water molecules diffuse through local field gradients during the echo time *TE*, then this will result in imperfect refocusing and hence to signal loss in

T_2-weighted spin-echo sequences (Thulborn *et al.* 1982; see also the ^{1}H NMR studies of red cell metabolism described by Brindle *et al.* 1979). Since water molecules can only diffuse a small distance in an echo time that may be of the order of 60 ms, this diffusion-dependent mechanism of signal loss should be more effective when the distances that characterize the gradients are small. Within the blood vessels themselves, these distances are indeed small, because the localization of deoxyhaemoglobin to red cells results in local field gradients in the periphery of each individual cell. In the adjacent tissue, however, the dimensions characterizing these local gradients should be determined by the dimensions of the vessels. Therefore, it can be anticipated that, in comparison with large draining vessels, small vessels such as capillaries should produce relatively large diffusion-dependent T_2 effects in the adjacent tissue. In principle, this might provide a means of distinguishing between the effects of large and small vessels, as discussed further in Section 4.6.4.

4.6.3 Recent observations

In the year after the initial Gd-DTPA study of Belliveau *et al.* (1991), several publications showed that T_2^*-weighted MRI could be used to map activated regions of the human brain without any requirement for extrinsic contrast agents (Kwong *et al.* 1992; Ogawa *et al.* 1992; Bandettini *et al.* 1992; Frahm *et al.* 1992; Blamire *et al.* 1992). Most of these studies involved mapping of the visual cortex, but some functional mapping with motor tasks was also described. In all of the studies, images were obtained under control and activated conditions, and comparison of one set of images with the other showed an increase in signal intensity within the activated regions, entirely consistent with the anticipated changes in venous oxygenation. However, it should be emphasized that there are a variety of mechanisms whereby haemodynamic changes can influence signal intensities, and a great deal more work is required to define the precise basis for all the changes that may be observed in activation studies. For example, changes in regional cerebral blood flow can influence signal intensities directly, and indeed the study of Kwong *et al.* (1992) showed that a T_1-sensitive sequence could also reveal activation, presumably as a result of increased flow during activation. Similarly, the EPISTAR technique discussed in Section 4.4 has recently been used to visualize activated regions of the brain through the detection of changes in regional blood flow (Edelman *et al.* 1994).

The study of the primary visual and motor cortices provides an appropriate means of establishing and validating this new method of functional neuroimaging, partly because these systems are relatively well characterized, and also because the haemodynamic changes (and hence signal intensity changes) associated with these visual and motor tasks are likely to be considerably greater than those associated with higher cognitive functions. One of the main requirements is to establish the extent to which such higher functions are accessible to investigation by MRI. Initial studies, for example observations involving word generation (Hinke *et al*. 1993; McCarthy *et al*. 1993; Rueckert *et al*. 1994), mental imagery (e.g imagining of motor and visual tasks; Rao *et al*. 1993, Le Bihan *et al*. 1993), and other cognitive tasks (see, for example, Kim *et al*. 1994) appear encouraging, as illustrated by the studies shown in Plate III. However, much more work still needs to be carried out in order to establish the sensitivity of MRI to the wide range of cognitive tasks that are of interest to the neuroscience community.

The initial functional MRI studies all used echo-planar imaging or magnetic field strengths of 2 T and above. Echo-planar imaging is desirable primarily because of its high-speed 'snapshot' capability, and because it permits wider coverage of the brain in a reasonable period of time; high fields are desirable because the susceptibility effects of deoxyhaemglobin become more pronounced as the field increases. However, at the time of these studies, neither echo-planar imaging nor field strengths of 2 T and above were widely available. While this situation will inevitably change, the question nevertheless arises as to how successfully functional MRI might be carried out using the large number of conventional MRI systems currently operating in routine clinical environments. It turns out that standard 1.5 T systems can indeed be used for at least some applications of functional imaging, including simple motor and visual task activation studies (see, for example, Connelly *et al*. 1993). This greatly facilitates the use of functional MRI for clinical as well as for the more neuroscientific applications.

One clinical area of interest relates to neurosurgical removal of lesions that are close to the primary sensory or motor cortex. It can often be difficult to predict the exact location of these primary cortical areas, due to normal biological variation and the distorting effects of the lesion itself. In order to permit maximal resection of such lesions, cortical mapping is performed. Until now, this has required direct electrical stimulation of the brain, often during an awake craniotomy. Using functional MRI, it is now possible to carry out non-invasive pre-surgical mapping of the sensory and motor cortex (Jack *et al*. 1994). The functional MRI maps can be validated by comparison with the standard intraoperative cortical mapping procedures and, once the MRI method has been extensively validated in this way, it could have major implications with respect to pre-operative planning and counselling of patients with lesions in these areas of the brain.

Functional MRI could also be of value in the investigation of patients with epilepsy; in fact, preliminary studies have shown that the technique can be used to map the cortical activation that occurs during focal seizures. In these studies, of a boy suffering from frequent partial motor seizures, functional MRI revealed sequential cortical activation in structurally abnormal regions of the brain (Jackson *et al*. 1994; Fig. 4.11). The activation was observed with each of five consecutive clinical seizures, and was also seen during a period

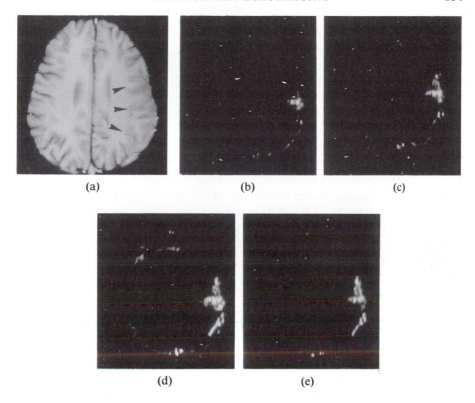

Fig. 4.11 (a) T_1-weighted image obtained from a 4-year-old child suffering from frequent partial motor seizures. The arrows indicate structurally abnormal regions of the brain. (b)–(e) are activation images of the same plane showing the time course and spatial distribution of the cortical activation associated with a clinical seizure. These images were obtained by subtracting baseline data (acquired between seizures) from data obtained during and around the time of the seizure. (b) represents data obtained just before the beginning of the clinical seizure, and shows that there was some activation before the seizure began; (c), (d) and (e) represent data obtained on onset of the seizure, 20 s, 60 s, and 100 s later. (Adapted from Jackson *et al.* (1994), *Neurology*, **44**, 850–6, with permission from *Neurology*.)

that was not associated with a detectable clinical seizure. This latter finding, together with the time-dependence of the signal changes observed, opens up the possibility of using functional MRI to detect not only the seizures themselves, but also sub-clinical or interictal events. The technique may therefore find relatively widespread application in the investigation of patients with intractable seizure disorders.

4.6.4 Technical considerations

While the initial experiences with functional MRI appear highly promising, it is nevertheless essential to bear in mind the possible limitations of the technique. One important point is that changes in signal intensity may be observed as a result of haemodynamic changes that are associated not only with capillaries, but also with larger draining vessels (see Lai *et al.* 1993). The draining vessels may be situated a long way 'downstream' of the activated tissue, and on task activation they may give raise to changes in signal intensity in regions that do not directly reflect the regions that are actually activated. For this reason, it is essential to establish, by one means or another, that the location of the observed signal changes does in fact reflect the focally activated regions. There are several ways in which this might be achieved. One is to combine functional MRI with angiographic techniques, so that the regions of increase in signal intensity can be compared directly with the anatomical location of the draining vessels. Alternatively, the signal from these vessels may be suppressed by using imaging sequences that give reduced signal from regions of high flow. In addition, multi-slice techniques that effectively give a three-dimensional activation pattern should help to establish the source of the signal increase. Finally, it has been pointed out that the combination of T_2-weighted and T_2^*-weighted imaging should help to distinguish between large and small vessels (see above), T_2-weighted and T_2^*-weighted effects becoming more similar as vessels become smaller (Ogawa *et al.* 1993; Weisskoff *et al.* 1994). In practice, there is a lot to be said for visualizing both the activated regions and the haemodynamic changes in the draining vessels, and the best approaches will no doubt preserve both types of information.

The effects of motion also need to be considered (Hajnal *et al.* 1994). If there is significant movement of the brain between the control and activated states, then the activation images will be heavily artefacted. However, while it may prove difficult to completely eliminate slight movements, it seems likely that, with appropriate experimental design and methods of image registration and analysis, it should be possible to cope with any resulting problems.

One of the potential attractions of functional neuroimaging with MRI is its high temporal resolution when compared with positron emission tomography. Echo-planar imaging in principle provides resolution of about 40 ms, but the question arises as to whether the temporal resolution is limited by the imaging sequence or by the time dependence of the physiological effects that are visualized by MRI. Numerous studies have shown that the increase in signal intensity on cortical activation occurs fairly gradually over several seconds, presumably reflecting the time constant of the haemodynamic events, and it may well be that this limits the achievable resolution. However, it seems possible that during the first tens of milliseconds of activation, the activated tissue may take up additional oxygen from the blood before there is a chance for the blood flow to increase. Under these circumstances, there may be an initial transient decrease in signal intensity on activation, followed by the increase that has formed the basis for all of the above studies. Indeed, initial studies using functional spectroscopy, in which the water signal is detected from a single selected region of interest, suggest that there may be a transient decrease in signal intensity which could form the basis for achieving improved temporal resolution (Ernst and Hennig 1994). With regard to spectroscopy, it is of interest to note that increases in lactate have been observed on task activation (see Shulman *et al.* 1993 and refs therein). The relatively poor temporal and spatial resolution associated with such metabolic studies imposes fairly severe limitations on the utility of this approach for functional studies. Nevertheless, metabolic studies of this type should shed further light on the control mechanisms underlying biochemical and physiological changes associated with task activation.

The question arises as to the sensitivity of functional neuroimaging, i.e. its ability to detect smaller activations than those so far observed. The percentage increase in signal intensity with visual and motor tasks is typically a few per cent, depending upon a number of technical factors, including the field strength at which the study is carried out. Many cognitive tasks of interest may be associated with only small haemodynamic changes, and hence with small percentage changes in signal intensity. The detection of signal changes of 1 per cent or less is feasible in principle, but will rely on high stability in machine performance, and on image registration and statistical methods of the type that have already been used for activation studies using positron emission tomography. One advantage of MRI is that tasks can be repeated many times, in single or multiple examinations, and the resulting signal-to-noise gains, together with improved image registration and statistical methods, should in due course enable small activations to be visualized reliably.

Finally, we need to consider the methods underlying the analysis and display of functional images. An 'activation image' can most simply be obtained by subtraction of a set of control images from a set obtained during the task. However, it is important to determine the statistical significance of the differences between the two data sets, and it is now common practice to display in the activation images only those pixels in which the differences achieve the required level of significance. Many studies involve cyclical control and activation periods, and the cyclical nature of the associated signal intensity changes provides additional information that can be exploited for the purposes of image analysis and display (Bandettini *et al.* 1993). This information can be incorporated into the statistical approaches that have been developed for activation studies using positron emission tomography (Friston *et al.* 1994). It is apparent that the future development of functional MRI will lean heavily on the use of such methods.

REFERENCES

Atkinson, D. J. and Edelman, R. R. (1991). Cineangiography of the heart in a single breath hold with a segmented TurboFLASH sequence. *Radiology*, **178**, 357–60.

Axel, L. and Dougherty, L. (1989). MR imaging of motion with spatial modulation of magnetization. *Radiology*, **171**, 841–5.

Balaban, R. S. and Cackler, T. L. (1992). Magnetization transfer contrast in magnetic resonance imaging. *Magn. Reson. Quart.* **8**, 116–37.

Bandettini, P. A., Wong, E. C., Hinks, R. S., Tikofsky, R. S., and Hyde, J. S. (1992). Time course EPI of human brain function during task activation. *Magn. Reson. Med.* **25**, 390–7.

Bandettini, P. A., Jesmanowicz, A., Wong, E. C., and Hyde, J. S. (1993). Processing strategies for time course data sets in functional MRI of the human brain. *Magn. Reson. Med.*, **30**, 161–73.

Belliveau, J. W., Kennedy, D. N., McKinstry, R. C., Buchbinder, B. R., Weisskopf, R. M., Cohen, M. S., *et al.* (1991). Functional mapping of the human visual cortex by magnetic resonance imaging. *Science*, **254**, 716–19.

Benveniste, H., Cofer, G. P., Piantadosi, C. A., Davis, J. N., and Johnson, G. A. (1992). Quantitative proton magnetic resonance imaging in focal cerebral ischemia in rat brain. *Stroke*, **22**, 259–68.

Blamire, A. M., Ogawa, S., Ugurbil, K., Rothman, D., McCarthy, G., Ellermann, J. M., *et al.* (1992). Dynamic mapping of the human visual cortex by high-speed magnetic resonance imaging. *Proc. Natl. Acad. Sci. USA*, **89**, 11069–73.

Bloch, F. (1946). Nuclear induction. *Phys. Rev.*, **70**, 460–74.

Bradley, W. G., jun. (1992). Flow phenomena. In *Magnetic resonance imaging* (eds D. D. Stark and W. G. Bradley, jun.), pp. 253–98. Mosby Year Book, St. Louis.

Brindle, K. M., Brown, F. F., Campbell, I. D., Grathwohl, C., and Kuchel, P. W. (1979). Application of spin-echo nuclear magnetic resonance to whole-cell systems. *Biochem. J.*, **180**, 37–44.

van Bruggen, N., Roberts, T. P .L. and Cremer, J. E. (1994). The application of magnetic resonance imaging to the study of experimental cerebral ischaemia. *Cerebrovasc. Brain Metab. Rev.*, **6**, 180–210.

Busza, A. L., Allen, K. L., King, M. D., van Bruggen, N., Williams, S. R., and Gadian, D. G. (1992). Diffusion-weighted imaging studies of cerebral ischemia in gerbils: potential relevance to energy failure. *Stroke*, **23**, 1602–12.

Connelly, A., Jackson, G. D., Frackowiak, R. S. J., Belliveau, J. W., Vargha-Khadem, F., and Gadian, D. G. (1993). Functional mapping of activated human primary cortex with a clinical MR imaging system. *Radiology*, **188**, 125–30.

Cowan, F. M., Pennock, J. M., Hanrahan, J. D., Manji, K. P., and Edwards, A. D. (1994). Early detection of cerebral infarction and hypoxic ischemic encephalopathy in neonates using diffusion-weighted magnetic resonance imaging. *Neuropediatrics*, **25**, 172–5.

Detre, J. A., Zhang, W., Roberts, D. A., Silva, A. C., Williams, D. S., Grandis, D. J., *et al.* (1994). Tissue perfusion specific imaging using arterial spin labeling. *NMR Biomed.*, **7**, 75–82.

Edelman, R. R., and Warach, S. (1993). Magnetic resonance imaging. *N. Engl. J. Med.*, **328**, 708–16, 785–91.

Edelman, R. R., Ahn, S., Chien, D., Li, W., Goldman, A., Mantello, M., *et al.* (1992).

Improved time-of-flight MR angiography of the brain with magnetization transfer contrast. *Radiology*, **184**, 395–9.

Edelman, R. R., Siewert, B., Darby, D. G., Thangaraj, V., Nobre, A. C., Mesulam, M.-M., *et al.* (1994). Qualitative mapping of cerebral blood flow and functional localization with echo-planar MR imaging and signal targeting with alternating radio frequency. *Radiology*, **192**, 513–20.

Eleff, S. M., Maruki, Y., Monsein, L. H., Traystman, R. J., Bryan, R. N., and Koehler, R. C. (1991). Sodium, ATP, and intracellular pH transients during reversible complete ischemia of dog cerebrum. *Stroke*, **22**, 233–41.

Ernst, T. and Hennig, J. (1994). Observation of a fast response in functional MR. *Magn. Reson. Med.*, **32**, 146–9.

Fuller, A. G., Howe, F. A., Hayes, C. E., Kliot, M., Winn, M. R., Bell, B. A., Griffiths, J. R., and Tsuruda, J. S. (1993). Magnetic resonance neurography. *Lancet*, **341**, 659–61.

Firmin, D. N., Nayler, G. L., Klipstein, R. H., Underwood, S. R., Rees, R. S. O., and Longmore, D. B. (1987). In vivo validation of MR velocity imaging. *J. Comput. Assist. Tomogr.*, **11**, 751–6.

Fox, P. T. and Raichle, M. E. (1986). Focal physiological uncoupling of cerebral blood flow and oxidative metabolism during somatosensory stimulation in human subjects. *Proc. Natl. Acad. Sci. USA*, **83**, 83, 1140–4.

Fox, P. T., Raichle, M. E., Mintun, M. A., and Dence, C. (1988). Nonoxidative glucose consumption during focal physiologic neural activity. *Science*, **241**, 462–4.

Frahm, J., Bruhn., H., Merboldt, K.-D., and Hanicke, W. (1992*a*). Dynamic MR imaging of human brain oxygenation during rest and photic stimulation. *J. Magn. Reson. Imaging*, **2**, 501–5.

Frahm, J., Gyngell, M. L., and Hanicke, W. (1992*b*). Rapid scan techniques. In *Magnetic resonance imaging* (eds. D. D. Stark, and W. G. Bradley, jun.), pp. 165–203. Mosby Year Book, St. Louis.

Friston, K. F., Jezzard, P., and Turner, R. (1994). The analysis of functional MRI time series. *Human Brain Mapping*, **1**, 153–71.

Haase, A., Frahm, J., Matthaei, D., Hanicke, W., and Merboldt, K.-D. (1986). FLASH imaging. Rapid NMR imaging using low flip-angle pulses. *J. Magn. Reson.*, **67**, 258–66.

Hajnal, J. V., Myers, R., Oatridge, A., Schwieso, J. E., Young, I. R., and Bydder, G. M. (1994). Artifacts due to stimulus correlated motion in functional imaging of the brain. *Magn. Reson. Med.*, **31**, 283–91.

Henkelman, R. M. and Bronskill, M. J. (1987). Artifacts in magnetic resonance imaging. *Rev. Magn. Reson. Med.*, **2**, 1–126.

Hinke, R. M., Hu, X., Stillman, A. E., Kim, S-G., Merkle, H., Salmi, R., *et al.* (1993). Magnetic resonance functional imaging of Broca's area during internal speech. *Neuro Report*, **4**, 675–8.

Jack, C. R., Thompson, R. M., Sharbrough, F. W., Kelly, P. J., Hanson, D. P., Butts, R. K., *et al.* (1994). Sensory motor cortex: correlation of presurgical mapping with functional MR imaging and invasive cortical mapping. *Radiology*, **190**, 1–8.

Jackson, G. D., Connelly, A., Cross, J. H., Gordon, I., and Gadian, D. G. (1994). Functional magnetic resonance imaging of focal seizures. *Neurology*, **44**, 850–6.

Kim S.-G., Ugurbil, K., and Strick, P. L. (1994). Activation of a cerebellar output nucleus during cognitive processing. *Science*, **265**, 949–51.

King, M. D., van Bruggen, N., Busza, A. L., Houseman, J., Williams, S. R., and Gadian, D. G. (1992). Perfusion and diffusion MR imaging. *Magn. Reson. Med.*, **24**, 288–301.

Kohno, K., Hoehn-Berlage, M., Mies, G., Back, T., and Hossmann, K. A. (1995). Relationship between diffusion-weighted MR images, cerebral blood flow, and energy state in experimental brain infarction. *Magn. Reson. Imaging*, **13**, 73–80.

Kondo, C., Caputo, G. R., Semelka, R., Foster, E., Shimakawa, A., and Higgins, C. B. (1991). Right and left ventricular stroke volume measurements with velocity-encoded cine MR imaging: *in vitro* and *in vivo* validation. *Am. J. Roentgenol.*, **157**, 9–16.

Kucharczyk, J., Roberts, T., Moseley, M. E., and Watson, A. (1993). Contrast-enhanced perfusion-sensitive MR imaging in the diagnosis of cerebrovascular disorders. *J. Magn. Reson. Imaging*, **3**, 241–5.

Kwong, K. K., Belliveau, J. W., Chesler, D. A., Goldberg, I. E., Weisskoff, R. M., Poncelet, B. P., *et al.* (1992). Dynamic magnetic resonance imaging of human brain activity during primary sensory stimulation. *Proc. Natl. Acad. Sci. USA*, **89**, 5675–9.

Lai, S., Hopkins, A. L., Haacke, E. M., Li, D., Wasserman, B. A., Buckley, P., *et al.* (1993). Identification of vascular structures as a major sourcer of signal contrast in high resolution 2D and 3D functional activation imaging of the motor cortex at 1.5 T: preliminary results. *Magn. Reson. Med.*, **30**, 387–92.

Le Bihan, D. and Turner, R. (1992). Diffusion and perfusion. In *Magnetic resonance imaging* (eds D. D. Stark, and W. G. Bradley, jun.), pp. 335–71. Mosby Year Book, St. Louis.

Le Bihan, D., Breton, E., Lallemand, D., Grenier, P., Cabanis, E., and Laval-Jeantet, M. (1986). MR imaging of intravoxel incoherent motions: application to diffusion and perfusion in neurologic disorders. *Radiology*, **161**, 401–7.

Le Bihan, D., Turner, R., Zeffiro, T. A., Cuenod, C. A., Jezzard, P., and Bonnerot, V. (1993). Activation of human primary visual cortex during visual recall: an MRI study. *Proc. Natl. Acad. Sci. USA*, **90**, 11802–5.

McCarthy, G., Blamire, A. M., Rothman, D. L., Gruetter, R., and Shulman, R. G. (1993). Echo-planar MRI studies of frontal cortex activation during word generation in humans. *Proc. Natl. Acad. Sci. USA.*, **90**, 4952–6.

Manning, W. J., Li, W., and Edelman, R. R. (1993). A preliminary report comparing magnetic resonance coronary angiography with conventional angiography. *N. Engl. J. Med.* **328**, 828–32.

Masaryk, T. J., Lewin, J. S., and Laub, G. (1992). Magnetic resonance angiography. In *Magnetic resonance imaging* (eds D. D. Stark, and W. G. Bradley, jun.), pp. 299–334. Mosby Year Book, St. Louis.

Mintorovich, J., Moseley, M. E., Chileuitt, L., Shimizu, H., Cohen, Y., and Weinstein, P. R. (1991). Comparison of diffusion- and T_2-weighted MRI for the early detection of cerebral ischemia and reperfusion in rats. *Magn. Reson. Med.*, **18**, 39–50.

Mohiaddin, R. H. and Longmore, D. B. (1993). Functional aspects of cardiovascular nuclear magnetic resonance imaging: techniques and application. *Circulation*, **88**, 264–81.

Moseley, M. E., Cohen, Y., Mintorovich, J., Chileuitt, L., Shimizu, H., Kucharczyk, J., *et al.* (1990). Early detection of regional cerebral ischemia in cats: comparison of diffusion- and T_2-weighted MRI and spectroscopy. *Magn. Reson. Med.*, **14**, 330–46.

Naritomi, H., Sasaki, M., Kanashiro, M., Kitani, M., and Sawada, T. (1988). Flow thresholds for cerebral energy disturbance and Na^+ pump failure as studied by ^{31}P and ^{23}Na NMR spectroscopy. *J. Cereb. Blood Flow Metab.*, **8**, 16–23.

Ogawa, S. and Lee, T-M. (1990*a*). Magnetic resonance imaging of blood vessels at high fields: *in vivo* and *in vitro* measurements and image simulation. *Magn. Reson. Med.*, **16**, 9–18.

Ogawa, S., Lee, T-M., Nayak, A. S., and Glynn, P. (1990*b*). Oxygenation-sensitive contrast in magnetic resonance image of rodent brain at high magnetic fields. *Magn. Reson. Med.*, **14**, 68–78.

Ogawa, S., Tank, D. W., Menon, R., Ellermann, J. M., Kim, S-G., Merkle, H. *et al.* (1992). Intrinsic signal changes accompanying sensory stimulation: functional brain mapping with magnetic resonance imaging. *Proc. Natl. Acad. Sci. USA*, **89**, 5952–5.

Ogawa, S., Menon, R. S., Tank, D. W., Kim, S.-G., Merkle, H., Ellermann, J. M., *et al.* (1993). Functional brain mapping by blood exygenation level-dependent contrast magnetic resonance imaging. A comparison of signal characteristics with a biophysical model. *Biophys. J.*, **64**, 803–12.

Pekar, J., Ligeti, L., Ruttner, Z., Sinnwell, T., Moonen, C. T. W., and McLaughlin, A. C. (1991). Combined ^{31}P/triple quantum filtered ^{23}Na NMR studies of ischemia in the cat brain. *Proc. Soc. Magn. Reson. Med.*, San Francisco meeting, p. 149.

Pennock, J. M., Cowan, F. M., Schweiso, J. E, Oatridge, A., Rutherfard, M. A., Pubowitz, L. M. S, and Bydder, G. M. (1994). Clinical role of diffusion-weighted imaging: neonatal studies. *MAGMA*. **2**, 273–8.

Perman, W. H. and Gado, M. H. (1992). *In. vivo* relationship between Gd-DTPA concentration and R2 for brain parenchyma and arterial blood. *Radiology*, **185**, 127.

Pike, G. B., Hu, B. S., Glover, G. H., and Enzmann, D. R. (1992). Magnetization transfer time-of-flight magnetic resonance angiography. *Magn. Reson. Med*, **25**, 372–9.

Rao, S. M., Binder, J. R., Bandettini, P. A., Hammeke, T. A., Yetkin, F. Z., Jesmanowicz, A., *et al.* (1993). Functional magnetic resonance imaging of complex human movements. *Neurology*, **43**, 2311–18.

Roberts, D., Detre, J. A., Bolinger, L., Insko, E. K., and Leigh, J. S., jun. (1994). Quantitative magnetic resonance iinaging of human brain perfusion at 1.5 T using steady-state inversion of arterial water. *Proc. Natl. Acad. Sci. USA*, **91**, 33–7.

Rosen, R. R., Belliveau, J. W., Buchbinder, B. R., McKinstrey, R. C., Porkka, L. M., Kennedy, D. N., *et al.* (1991). Contrast agents and cerebral hemodynamics. *Magn. Reson. Med.*, **19**, 285–92.

Rucckert, L., Appollonio, I., Grafmaii, J., Jezzard, P., Johnson, R., jun., Le Bihan, D., and Turner, R. (1994). MRI functional activation of the left frontal cortex during covert word production. *J. Neuroimaging*, **4**, 67–70.

Sakuma, H., Fujita, N., Koo, T. K. F., Caputo, G. R., Nelson, S. J., Hartiala, J., *et al.* (1993). Evaluation of left ventricular volume and mass with breath-hold cine MR imaging. *Radiology*, **188**, 377–80.

Shulman, R. G., Blamire, A. M., Rothman, D. L., and McCarthy, G. (1993). Nuclear magnetic resonance imaping and spectroscopy of human brain function. *Proc. Natl. Acad. Sci, USA*, **90**, 3127–33.

Stark, D. D. and Bradley, W. G, jun. (1992). *Magnetic resonance imaging*. Mosby Year Book, St. Louis.

Stehling, M. K., Turner, R., and Mansfield, P. (1991). Echo-planar imaging: magnetic resonance imaging in a fraction of a second. *Science*, **254**, 43–50.

Steinberg, E. P. (1993). Editorial. Magnetic resonance coronary angiography — assessing an emerging technology. *N. Engl. J. Med.*, **328**, 879–80.

Stejskal, E. O. and Tanner, J. E. (1965). Spin diffusion measurements: spin-echoes in the presence of a time-dependent field gradient. *J. Chem. Phys.*, **42**, 288–92.

Thulborn, K. R., Waterton, J. C., Matthews, P. M., and Radda, G. K. (1982).

Oxygenation dependence of the transverse relaxation time of water protons in whole blood at high field. *Biochim. Biophys. Acta*, **714**, 265–70.

Turner, R. and Jezzard, P. (1994). How to see the mind. *Physics World*, **7(8)**, 29–33.

Turner, R., Le Bihan, D., Moonen, C. T. W., Despres, D., and Frank, J. (1991). Echo-planar time course MRI of cat brain oxygenation changes. *Magn. Reson. Med.*, **22**, 159–66.

Verheul, H. B., Balazs, R., Berkelbach van der Sprenkel, J. W., Tulleken, C. A, F., Nicolay, K., Tamminga, K. S., *et al.* (1994). Comparison of diffusion-weighted MRI with changes in cell volume in a rat model of brain injury. *NMR Biomed.*, **7**, 96–100.

Warach, S., Chien, D., Li, W., Ronthal, M., and Edelman, R. R. (1992). Fast magnetic resonance diffusion-weighted imaging of acute human stroke. *Neurology*, **42**, 1717–23.

Warach, S., Gaa, J., Siewert, B., Wielopolski, P., and Edelman, R. (1995). Acute human stroke studied by whole brain echo planar diffusion-weighted MRI. *Ann. Neurol.*, **37**, 231–41.

Watson, A. D., Rocklage, S. M., and Carvlin, M. J. (1992). Contrast agents. In *Magnetic resonance imaging* (eds D. D. Stark, and W. G. Bradley, jun.), pp. 372–437. Mosby Year Book, St. Louis.

Weisskoff, R. M., Chesler, D., Boxermanm, J. L., and Rosen, B. R. (1993). Pitfalls in MR measurement of tissue blood flow with intravascular tracers: which mean transit time? *Magn. Reson. Med.*, **29**, 553–9.

Weisskoff, R. M., Zuo, C. S., Boxerman, J. L., and Rosen, B. R. (1994). Microscopic susceptibility variation and transverse relaxation: theory and experiment. *Magn. Reson. Med.*, **31**, 601–10.

Wentz, K. U., Rother, J., Schwartz, A., Mattle, H. P., Suchalla, R., and Edelman, R. R. (1994). Intracranial vertebrobasilar system: MR angiography. *Radiology*, **190**, 105–10.

White, S. J., Hajnal, J. V., Young, I. R., and Bydder, G. M. (1992). Use of fluid attenuated inversion recovery pulse sequences for imaging the spinal cord. *Magn. Reson. Med.*, **28**, 153–62.

Wilke, N., Simm, C., Zhang, J., Ellermann, J., Ya, X., Merkle, H., *et al.* (1993). Contrast-enhanced first pass myocardial perfusion imaging: correlation between myocardial blood flow in dogs at rest and during hyperemia. *Magn. Reson. Med.*, **29**, 485–97.

Wilke, N., Kroll, K., Merkle, H., Wang, Y., Ishibashi, Y., Xu, Y. *et al.* (1995). Regional myocardial blood volume and flow: first-pass MR imaging with polylysine-Gd-DTPA. *J. Magn. Reson. Imaging*, **5**, 227–37.

Wood, M. L. and Ehman, R. L. (1992). Effects of motion in MR imaging. In *Magnetic resonance imaging* (eds D. D. Stark and W. G. Bradley, jun., pp. 145–64. Mosby Year Book, St. Louis.

Zhang, W., Williams, D. S., and Koretsky, A. P. (1993). Measurement of rat brain perfusion by NMR using spin labeling of arterial water: *in vivo* determination of the degree of spin labeling. *Magn. Reson. Med.*, **29**, 416–21.

Zhong, J., Petroff, O. A. C., Prichard, J. W., and Gore, J. C. (1993). Changes in water diffusion and relaxation properties of rat cerebrum during status epilepticus. *Magn. Reson. Med.*, **30**, 241–6.

Zhong, J., Petroff, O. A .C., Prichard, J. W., and Gore, J. C. (1995). Barbiturate-reversible reduction of water diffusion coefficient in fluorthyl-induced status epilepticus in rats. *Magn. Reson. Med.*, **33**, 253–6.

5

The theoretical basis of NMR

5.1 INTRODUCTION

In order to understand how radiation is absorbed by matter, we need to recognize firstly that radiation is quantized and secondly that atoms and molecules can only have certain discrete energy levels. These are concepts that are outside the realm of classical physics, and therefore a rigorous treatment of any spectroscopic technique requires the use of the branch of physics that is termed quantum mechanics. However, certain aspects of NMR can be understood surprisingly adequately and clearly in terms of classical physics. For this reason descriptions of NMR theory frequently contain a mixture of quantum mechanical and classical treatments, depending on which provides the simpler or more helpful picture of what is happening. In this chapter, we shall first give a brief quantum mechanical description of the basic properties of atomic nuclei, explaining how magnetic nuclei interact with applied magnetic fields. The treatment that is presented here is intended to be comprehensible to readers with no prior knowledge of quantum mechanics; however, some of the results that are given will have to be taken on trust, for the theory on which they are based is beyond the scope of a book of this type. We shall then proceed to describe NMR in terms of the classical model that is particularly well suited to explaining many of the practical aspects of NMR, such as the effects of radiofrequency pulses. The latter part of the chapter is concerned with the Fourier transform, which plays a critical role in NMR imaging and spectroscopy.

We begin by noting that certain atomic nuclei, such as ^{1}H, ^{13}C, and ^{31}P, possess a property known as spin, which can be visualized as a spinning motion of the nucleus about its own axis. These nuclei, in common with other spinning objects, possess the property of angular momentum, and they also, by virtue of their electrical charge, have magnetic properties. This nuclear magnetism is analogous to the magnetism generated by an electrical current circulating in a small loop of wire; such a current loop behaves like a small bar magnet and, similarly, the charged, spinning nucleus can be regarded as a tiny bar magnet, rotating about its own axis (see Fig. 1.3(a)).

The expression 'magnetic dipole moment' (or more simply 'magnetic moment') is often used to describe the properties of magnetic objects; it defines the turning moment that is experienced by the object when it is placed in an applied magnetic field, and this is the term that is used to describe the properties of magnetic nuclei. The angular momentum and magnetic moment are closely related to each other, and it is for this reason that we start the quantum mechanical section with a brief discussion of the quantized nature of the nuclear angular momentum.

5.2 QUANTUM MECHANICAL DESCRIPTION

5.2.1 The atomic nucleus

Quantum mechanics tells us that the angular momentum of atomic nuclei can only have certain discrete values, specified by a quantum number I. The magnitude p of the angular momentum is given by

$$p = \hbar I (I + 1)\}^{1/2} \qquad (5.1)$$

where $\hbar$ is equal to $h/2\pi$, h being the Planck constant. The quantum number I, usually called the spin of the nucleus, may only have-integral or half-integral values as follows.

1. I is integral for nuclei with even mass number.
2. I is zero for nuclei with even numbers of both neutrons and protons. This is because of the tendency for both neutrons and protons to form pairs in such a way that the individual spins cancel out. Therefore, ^{12}C and ^{16}O have zero spin and do not produce NMR signals.
3. I is half-integral for nuclei with odd mass numbers. Nuclei of spin $\frac{1}{2}$, such as ^{1}H, ^{13}C, and ^{13}P, are particularly important in NMR, as these are the nuclei that tend to have the most appropriate NMR characteristics.

Angular momentum is a vector property, i.e. it is specified by both magnitude and direction. In order to specify the direction of the angular momentum, it is necessary to introduce a second quantum number m, for it is found that the angular momentum vector can only have certain discrete orientations with respect to any given direction (e.g. the z-direction). The component p_z of angular momentum along the z-direction is given by

$$p_z = m\hbar \qquad (5.2)$$

m may have any of the $2I + 1$ values, $I, I - 1, \ldots -I$, and so for a nucleus of spin $\frac{1}{2}$, m can be $+\frac{1}{2}$ or $-\frac{1}{2}$. Therefore, for such a nucleus

$$p_z = \pm \tfrac{1}{2}\hbar \qquad (5.3)$$

and these two spin states are illustrated in Fig. 5.1.

As mentioned above, the magnetic moment of a nucleus is closely related to its angular momentum; in fact, the magnetic moment has the same direction as the angular momentum, and has magnitude μ given by

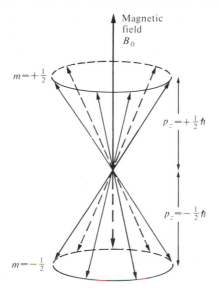

Fig. 5.1 Diagram illustrating the allowed orientations of the angular momentum vector of a nucleus of spin $\frac{1}{2}$. The orientations are specified by the quantum number m, and describe two cones.

$$\mu = \gamma p \qquad (5.4)$$

where γ is a proportionality constant known as the gyromagnetic (or magnetogyric) ratio of the nucleus. Thus, the component of the magnetic dipole moment along the z-axis can be written

$$\mu_z = \gamma \hbar m \qquad (5.5)$$

It should be noted that the magnitude of γ, and hence of the nuclear magnetic moment, cannot be predicted from classical physics. For most nuclei γ is positive, but for some, such as ^{15}N, γ is negative.

5.2.2 The interaction of the nucleus with a static magnetic field

If a static magnetic field[1] B_0 is applied along the z-axis, the nucleus acquires energy E as a result of the interaction between the field B_0 and the nuclear moment, the magnitude of which is given by

$$E = -\mu_z B_0$$

or, from eqn (5.5)

[1] A magnetic field can be described in terms of its flux density B_0 which determines the force exerted on other magnets. In this book we refer to B_0 for simplicity as the magnetic field rather than the magnetic flux density.

$$E = -\gamma \hbar m B_0 \qquad (5.6)$$

Since m can have any of the values $I, I-1, \ldots, -I$, the nuclear energy levels are split into $2I+1$ states by the application of the field, as was shown in Fig. 1.4 for a nucleus of spin $I = \frac{1}{2}$. As adjacent energy states differ by 1 in their value of m, the energy difference ΔE between adjacent states is given by

$$\Delta E = \gamma \hbar B_0 \qquad (5.7)$$

5.2.3 The effect of an oscillating magnetic field

In order to obtain NMR signals, it is necessary to induce transitions between the different energy states. Transitions between adjacent states are induced by the application of a suitable oscillating magnetic field B_1 in the xy-plane. The field must oscillate with a frequency ν_0 that satisfies the fundamental relationship of spectroscopy (see Section 1.2)

$$\Delta E = h\nu_0$$

Therefore, using eqn (5.7) we find that

$$\nu_0 = \gamma B_0 / 2\pi \qquad (5.8)$$

If we write $\omega_0 = 2\pi\nu_0$, where ω_0 is known as the angular frequency, then

$$\omega_0 = \gamma B_0 \qquad (5.9)$$

Equation (5.8), or eqn (5.9), expresses the resonance condition for NMR. Note that only transitions between adjacent states take place. This is an example of the quantum mechanical 'selection rules' that govern transitions between energy levels. Since γ differs for each nuclear isotope, different nuclei resonate in a given field B_0 at widely different frequencies. At conventional values of B_0 the frequencies occur in a convenient radiofrequency band, and the oscillating field B_1 is commonly referred to as the radiofrequency (r.f.) field.

5.2.4 The populations of the nuclear energy states

The populations of nuclei in the various energy states (see Fig. 5.2) are determined by the Boltzmann distribution; this specifies that at a thermal equilibrium characteristic of the temperature T the relative numbers n^+ and n^- of nuclei in the spin $+\frac{1}{2}$ and $-\frac{1}{2}$ states are given by

$$n^-/n^+ = \exp(-\Delta E/kT)$$
$$= \exp(-\gamma \hbar B_0/kT) \qquad (5.10)$$

where k is the Boltzmann constant. For protons in a magnetic field of 5 tesla the magnitude of ΔE is only 10^{-6} electron volts (eV), whereas at $20\,°C$ the thermal energy kT is about 2.5×10^{-2} eV. Therefore, $\exp(-\Delta E/kT)$ is very

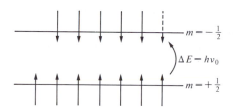

Fig. 5.2 Nuclei of spin $I = \frac{1}{2}$ in their two possible spin states. The nuclei of spin $m = +\frac{1}{2}$ (upward-pointing arrows) have lower energy then those of spin $m = -\frac{1}{2}$ (downward-pointing arrows); the $m = +\frac{1}{2}$ state is therefore the slightly more populated of the two states. A transition from the lower to the higher state is indicated by the curved arrow; transitions such as this tend to equalize the populations and hence lead to saturation (see Section 7.3.2).

close to unity, and the populations differ by less than 1 part in 10^4. Absorption of energy by the nuclei from the oscillating magnetic field B_1 relies on there being a population difference between adjacent states; if the populations were equal there would be equal numbers of transitions in both directions, resulting in no net absorption of energy and no signal. The small energy difference between the states therefore leads to a very weak absorption of energy and is responsible for the inherently low sensitivity of NMR. Note that an increase in the magnetic field B_0 increases the energy difference between adjacent states (eqn (5.7)) and hence their population difference (eqn (5.10)), and therefore considerably enhances the net absorption of energy. As a result, a high magnetic field is generally desirable to improve the signal-to-noise ratio.

There are two aspects of this basic quantum mechanical treatment that are consistent with the adequacy of the classical picture of NMR. First, the Planck constant is absent from eqn (5.9). Secondly, the predictions of quantum theory and classical physics differ considerably for systems in which the separation, ΔE, of the energy levels is much greater than the mean thermal energy kT. However, they tend to converge when ΔE is much less than kT, which is true in the case of NMR. We now move on to the classical description of NMR to continue the picture of how NMR signals are excited and detected.

5.3. CLASSICAL DESCRIPTION OF NMR

If a bar magnet is placed in a magnetic field it aligns itself parallel to the field, because this is the orientation of least energy. However, if the magnet possesses angular momentum it will not align parallel to the field, but instead will precess about the field with a characteristic angular frequency ω_0 (see Fig. 5.3). (Angular frequency is the rate of rotation in radians per second.) This behaviour is analogous to that of a spinning gyroscope in the earth's

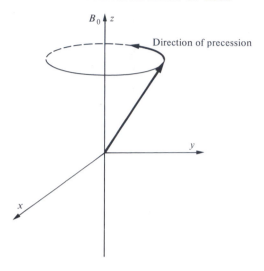

Fig. 5.3 A magnet possessing angular momentum precesses in an applied field B_0. The heavy arrow indicates the orientation of the magnet, and of its angular momentum.

gravitational field. If we regard the nucleus as a tiny bar magnet possessing angular momentum, it is possible to show, using classical physics, that the nuclear magnet, or moment, precesses about the direction of an applied field B_0 with angular frequency $\omega_0 = \gamma B_0$. This is known as the Larmor frequency and is identical to the resonance frequency derived from quantum theory (see eqn (5.9)).

In NMR experiments we do not of course study individual nuclei, but rather a sample containing a large number (typically 10^{18} or more) of magnetic nuclei. These nuclear magnets all precess about B_0 in the same direction, regardless of the value of their quantum number m. Since there is no preferred orientation in the plane perpendicular to B_0 (the xy-plane), the net component of magnetic moment in the xy-plane is zero. However, there is a net magnetization (defined as magnetic moment per unit volume) along the z-axis (see Fig. 5.4) because there are slightly more nuclei oriented with the field than against it, i.e. there are more nuclei in the lower energy state than in the upper state.

In order to detect the magnetization set up within a sample by the B_0 field, it is necessary to tilt the magnetization towards or into the xy-plane. It turns out that this can be accomplished by means of a radiofrequency field applied in the xy-plane. In order to understand how this process works, it is important to appreciate the concept of the rotating frame of reference.

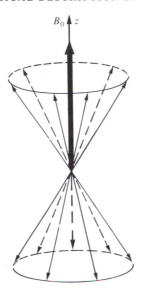

Fig. 5.4 In the presence of a field B_0, there is a net magnetization along the direction of the field, i.e. along the z-axis.

5.3.1 The rotating frame of reference and the B_1 field

We have seen that the application of a static magnetic field B_0 causes the nuclear magnets to precess about the direction of B_0 with angular frequency $\omega_0 = \gamma B_0$. Let us consider a world rotating with angular frequency ω_r relative to the laboratory, i.e. relative to the sample and NMR system (see Fig. 5.5). In this so-called rotating frame of reference the nuclear magnets appear to precess with angular frequency $\omega_0 - \omega_r$. If ω_r happens to be equal to ω_0, then in this particular frame of reference $\omega_0 - \omega_r$ is equal to zero, i.e. the nuclear magnets do not appear to precess at all, and by analogy with eqn (5.9) the apparent static magnetic field must also be zero. We shall now imagine ourselves to be in this rotating frame of reference for which $\omega_0 = \omega_r$. By analogy with the fixed laboratory frame of reference, which has coordinates x, y, and z, we label the coordinates of the rotating frame x', y', and z' (in fact z and z' are equivalent).

Let us apply, along the x'-axis of the rotating frame of reference, a field B_1, which is static in this frame. Then, *just as the application of B_0 causes the nuclei to precess about its direction with angular frequency γB_0, so the application of B_1 in the rotating frame causes the nuclear magnetization to precess about B_1 with angular frequency γB_1.* This is because the B_1 field is the only apparent field experienced in the rotating frame. If the field B_1 is

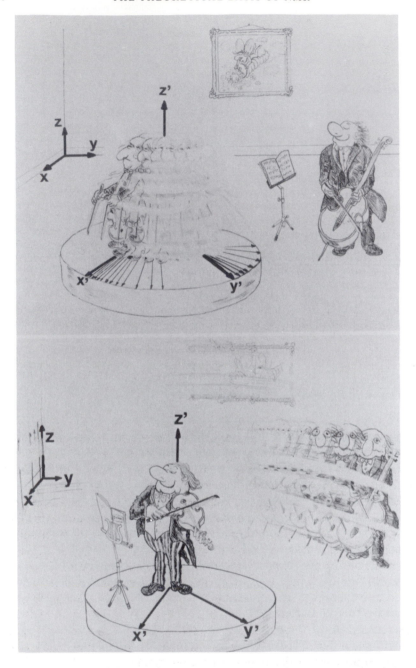

Fig. 5.5 Illustration of the rotating frame of reference. (Reprinted with permission of Springer-Verlag GmbH & Co. & K G from Roth (1984), NMR-tomographie und-spektroscopie in der Medizin.)

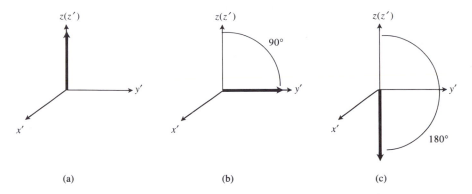

Fig. 5.6 (a) Prior to the application of a radiofrequency pulse, the *net* component of magnetization, indicated by the heavy arrow, is along the z-axis. (b) The effect of a 90° pulse applied along the x'-axis is to tilt the magnetization on to the y'-axis. (c) Similarly, a 180° pulse tilts the magnetization through 180°, i.e. it inverts the magnetization.

applied for time t_p, the nuclei will rotate through an angle θ given by the product of angular frequency with time, i.e.

$$\theta = \gamma B_1 t_p \tag{5.11}$$

The nuclei will rotate through an angle of 90° (corresponding to $\theta = \pi/2$ radians) if t_p is such that

$$\gamma B_1 t_p = \pi/2 \tag{5.12}$$

A pulse of B_1 field that has this duration is known as a 90° pulse, and its effect is to tilt the net magnetization away from the z- (or z'-) axis into the x'y'-plane of the rotating frame (see Fig. 5.6). Similarly, a 180° pulse (for which $\gamma B_1 t_p = \pi$) tilts the magnetization on to the negative z-axis.

Let us now consider the nature of the field B_1, which is static in the rotating frame. Reverting to the normal fixed frame, we see that in this frame B_1 must correspond to a field that rotates about the z-axis with angular frequency ω_0. The simplest method of generating this rotating B_1 field is to apply an oscillating field along a given direction (e.g. the x-direction) in the xy-plane, for such a field can be broken down into two components rotating in opposite directions. This is analogous to the breaking down of plane-polarized light into left- and right-handed circularly polarized light. The component rotating in the opposite direction to the nuclear precession can be shown to have a negligible effect in NMR, and so the oscillating B_1 field is effectively equivalent to the required rotating field.

We therefore find that macroscopic changes in the nuclear magnetization can be induced by applying an oscillating B_1 field that has the same frequency as the Larmor frequency of the nuclei. This classical result agrees perfectly with the quantum mechanical result which states that resonance is

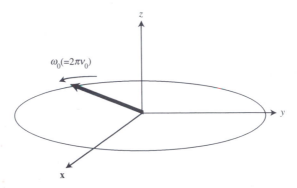

Fig. 5.7 Magnetization M_{xy} rotating about the z-axis with angular frequency ω_0.

achieved by application of an oscillating magnetic field of frequency ω_0 in the xy-plane. The equality of the Larmor frequency and the frequency of the B_1 field is analogous to other resonance phenomena for which the 'driving' frequency is equal to a characteristic frequency of the system. The B_1 field is in fact generated simply by passage of an oscillating electrical current through the transmitter coil.

5.4 THE BASIS OF SIGNAL DETECTION

In the absence of an applied B_1 field the nuclear magnets precess randomly about the field B_0 at their characteristic Larmor frequency. At any instant the net component of magnetization in any direction within the xy-plane is zero and no signal is observed. Now consider the effect of a 90° pulse of radiofrequency field B_1, which tilts the net nuclear magnetization M_0 away from the z-axis into the $x'y'$-plane of the rotating frame, and therefore into the xy-plane of the fixed frame. Following the pulse, the nuclear spins experience only the static field B_0 and so they continue to precess about B_0. However, now their precession is *not* totally random; a net component of magnetization M_{xy} has been generated in the xy-plane. This component M_{xy} rotates *coherently* about B_0 at the frequency ω_0 (Fig. 5.7) and induces an electromotive force (e.m.f.) in the receiver coil. The induced e.m.f. oscillates at the frequency ω_0 and its magnitude is governed by Faraday's law of magnetic induction. This e.m.f. is then amplified and processed on the receiving side of the NMR instrument to give a recognizable NMR signal. In practice the same radiofrequency coil can be used for transmitting the B_1 field and for receiving the resulting signal. Alternatively, separate coils can be used for transmission and reception, as discussed in Section 7.4.

5.5 THE FREE INDUCTION DECAY

We now move on to consider the equations describing the coherent rotation of M_{xy} following a 90° pulse. If M_{xy} is oriented along the x-axis at time $t = 0$, then, in the absence of any relaxation,

$$M_x = M_0 \cos \omega_0 t \qquad (5.13)$$

$$M_y = M_0 \sin \omega_0 t \qquad (5.14)$$

where at any time M_x and M_y are the components of magnetization along the x- and y-axes respectively.

Following a radiofrequency pulse, the magnetization M_{xy} actually decays with a time constant T_2^*, so the above equations should be modified to read

$$M_x = M_0 \cos \omega_0 t e^{-t/T_2^*} \qquad (5.15)$$

$$M_y = M_0 \sin \omega_0 t e^{-t/T_2^*} \qquad (5.16)$$

These decaying waveforms are represented in Fig. 5.8. The e.m.f. induced in the receiver coil represents one of these components of magnetization,[1] giving a signal that is commonly referred to as a free induction decay (FID); the expression free induction decay was introduced in the early days of NMR as a description of the decay of the induced signal arising from free precession of the nuclei in the field B_0.

More generally, a free induction decay contains many frequency components; it then has a much more complex appearance, and is very difficult to interpret visually. In order to transform this decaying signal (or a series of such signals) into an understandable form, it is necessary to apply the mathematical manipulation known as Fourier transformation, which converts a time-dependent signal into its equivalent frequency components, thereby generating a recognizable spectrum or image. The process of Fourier transformation is discussed in some detail in Section 5.7, but first it is important to introduce the concept of phase.

5.6 PHASE

Consider magnetization M_{xy} rotating about the z-axis with angular frequency $\omega_0 = 2\pi\nu_0$ as shown in Fig. 5.7. The magnetization completes one cycle (i.e. it rotates through 360°, or 2π radians) ν_0 times per second; hence it rotates through 360° or 2π radians in a time interval $1/\nu_0$, through 90° or $\pi/2$ radians in the time $1/4\nu_0$, and more generally through an angle of ϕ radians

---

[1] The use of quadrature detection coils (see Section 7.4.6) effectively exploits both of these components.

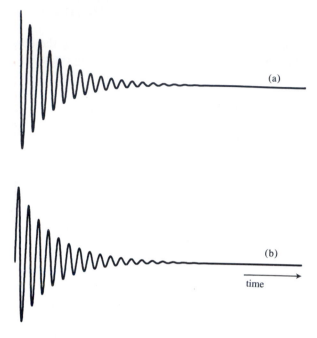

Fig. 5.8 Two free induction decays, 90° out of phase with each other. Note that in (a) the signal has its maximum value at zero time, whereas in (b) the signal is zero at zero time.

in time $t = \phi/2\pi\nu_0$. ϕ is commonly termed the phase angle, or more simply the phase of the magnetization. Thus in time t, the magnetization undergoes a phase change

$$\phi = 2\pi\nu_0 t \tag{5.17}$$

or

$$\phi = \omega_0 t \tag{5.18}$$

In magnetic resonance it is commonly important to consider phase *differences* between signals, rather than absolute phase angles. If two nuclei have resonance frequencies that differ by $\Delta\nu_0$, then following a 90° pulse they will precess at different frequencies, and after a time t they will be phase shifted from each other by an angle $\phi = 2\pi\Delta\nu_0 t$, as shown in Fig. 5.9. Such phase shifts are readily produced by the use of field gradients, which cause nuclei in different locations to precess at different frequencies; these shifts form the basis of phase encoding in MRI, and more specifically can be exploited as a means of generating contrast, for example in one form of magnetic resonance angiography. The phase of signals can also be manipulated and exploited by

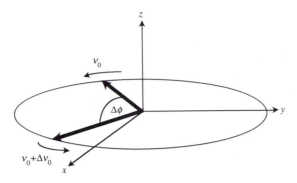

Fig. 5.9 Two nuclei with resonance frequencies that differ by Δv_0 precess at different frequencies and after a time t they are phase-shifted from each other by an angle
$$\Delta\phi = 2\pi\Delta v_0 t.$$

varying the phase and timing of radiofrequency pulses; for example, 90° phase shifts are frequently exploited in pulse sequences (see, for example, the Carr–Purcell–Meiboom–Gill sequence, which is discussed in Section 8.3).

We now move on to discuss the process of Fourier transformation, which provides a means of disentangling the phase information, as well as the amplitudes and frequencies of the various signals in a free induction decay.

5.7 FOURIER-TRANSFORM NMR

Imaging and spectroscopy both depend upon the detection of signals covering a range of frequencies; in imaging the frequencies give spatial information, while in spectroscopy they give chemical information. In principle, the signals could be sampled one by one, as in the early days of NMR, when spectra were obtained in the continuous-wave mode. In this method, the B_1 field was applied continuously in the xy-plane, and the B_0 field was swept through a range of field strengths in order to obtain a spectrum in which signal amplitude was plotted as a function of field strength. However, this was an inefficient process, in that only a fraction of the total collection time was spent observing any single component of the spectrum. It would be very advantageous if a technique could be developed whereby all the signals could be detected simultaneously. The uses of pulses of radiofrequency field were recognized early in the development of NMR, but pulsed NMR was little used in chemical analysis for many years. Then in the mid-1960s Ernst and Anderson showed that by use of Fourier transformation the pulsed approach could give high-resolution spectra equivalent to those obtained in continuous-wave NMR. Pulsed NMR has the great advantage that all signals can be excited simultaneously, and before

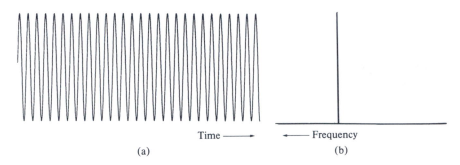

Time ⟶   ⟵ Frequency
(a)   (b)

Fig. 5.10 A wave can be expressed as a function of time (a) or of frequency (b). (a) represents a waveform that continues for ever.

too long, pulsed Fourier-transform NMR instruments took over from the old continuous-wave systems. The studies of living systems that have taken place since the early 1970s have been critically dependent upon Fourier-transform methods, and the remainder of this chapter is concerned with the Fourier transform and its roles in both imaging and spectroscopy.

The initial use of Fourier-transform NMR was in spectroscopy rather than imaging, and conceptually its use in spectroscopy provides a somewhat more gentle introduction, if only because spectroscopy can be carried out using one frequency dimension whereas imaging normally requires a two-dimensional analysis. We therefore begin our discussion of the Fourier transform by reference to spectroscopy, before moving on to its use in imaging.

5.7.1 The Fourier transform in MRS

As discussed above, the signal that is observed in response to a radiofrequency pulse, or a series of pulses, does not look like a spectrum or an image; in its simplest form it may appear as an oscillation that gradually decays away. In order to obtain a conventional spectrum in which signal amplitude is plotted as a function of frequency, this response is subjected to Fourier transformation. This mathematical procedure enables a quantity that varies with time to be analysed in terms of the sum of its frequency components. In this section we shall present a qualitative pictorial consideration of the way in which signals can be represented as a function of either time or frequency. A more mathematical description of the Fourier transform is given in the Appendix to this chapter (p. 164).

Let us consider the analogy of a tuning fork. When it is struck it responds by ringing with a characteristic frequency. This ringing can be regarded as a wave of the form shown in Fig. 5.10(a). Alternatively, the sound can be expressed in terms of its frequency as shown in Fig. 5.10(b) where the amplitude of the response is plotted as a function of frequency. Figure 5.10(b) expresses

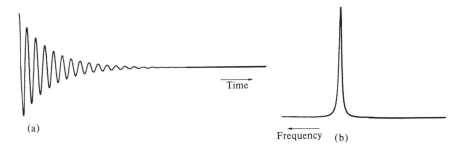

Fig. 5.11 A waveform expressed as a function of time (a) and of frequency (b).

the perfect quality of the note in the sense that it contains one single frequency component. A single frequency component corresponds to a perfect sine or cosine wave that continues for ever, and so the long lifetime of the response as shown in Fig. 5.10(a) equally expresses the perfect quality of the note.

Suppose that the note is less ideal in that it contains a distribution of frequencies centred about v_0 as shown in Fig. 5.11(b). Let us now consider the nature of the ringing. Initially, the response has a large amplitude which represents the fact that all frequency components start off in phase with each other. However, because the frequency components all oscillate at slightly different frequencies, they will gradually become out of phase with each other and destructive interference will take place. The amplitude of the ringing will gradually decay to zero as shown in Fig. 5.11(a). If the frequency distribution were greater, the decay would take place more rapidly; there is an inverse relationship between the length of time for which the ringing takes place and the spread in frequency. The two representations shown in Fig. 5.11 (and in Fig. 5.10) are equivalent;[1] the one specifies the other and they are connected to each other by the mathematical device of Fourier transformation. If the decay of the ringing is exponential, then its time constant T_c is related to the width at half-height of the frequency response by the relationship

$$1/T_c = \pi\Delta v_{1/2} \qquad\qquad (5.19)$$

The similarity between this equation and eqns (1.9) and (1.10) should be noted.

If several different tuning forks were struck simultaneously, the response expressed as a function of time would be complicated, as shown in Fig. 5.12(a). Nevertheless, the human ear would be capable of discriminating between the constituent frequencies, and indeed a musical person could identify the precise frequencies that were present. Effectively, the person would be Fourier transforming a complicated wave-pattern to give the frequency distribution shown in Fig. 5.12(b).

[1] Provided that phase information is preserved in the frequency response.

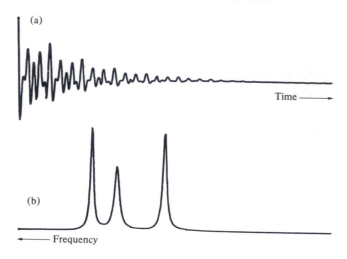

Fig. 5.12 A waveform expressed as a function of time (a) and of frequency (b).

Now, consider a wave of the form shown in Fig. 5.13(a). This is a wave of frequency v_0 which is cut off after a time t. Suppose we wish to measure the frequency. We could imagine a device that does this by counting the number of maxima in the waveform in the time t. In the example shown, where $t = 1$ s, the frequency would be measured at 11 waves per second, i.e. 11 Hz. However, our device might equally measure the number of minima, in which case the frequency would appear to be 10 Hz. There is an inherent uncertainty in the measured frequency which corresponds to approximately one wave in the time interval of the measurement, and this is true regardless of the duration of the time interval. Therefore, if the frequency of a waveform is measured over a time period t, there is necessarily an uncertainty in the measured frequency of 1 wave in t seconds, i.e. $1/t$ Hz. This is equivalent to there being an effective spread in frequency $\Delta v \approx 1/t$. Such a result can be obtained rigorously by Fourier analysis, and the frequency distribution corresponding to the pulse is shown in Fig. 5.13(b). The expression $\Delta v \approx 1/t$ is equivalent to one of the uncertainty principle relationships of quantum mechanics and is similar in form to eqn (5.19). The physical meaning of Fig. 5.13 is that the pulse shown in Fig. 5.13(a) could be generated by summing a distribution of waves of infinite duration whose frequencies and amplitudes are given by Fig. 5.13(b). Since $\Delta v \approx 1/t$, the frequency spread increases as t decreases.

Let us now return to the detection of NMR signals. We wish to optimize the sensitivity of the technique by simultaneously exciting all of the nuclei, the resonance frequencies of which may extend over a range of say, 5 kHz. We now see that we should be able to do this by applying a pulse of radiofrequency field, the duration of which is sufficiently short for the effective frequency

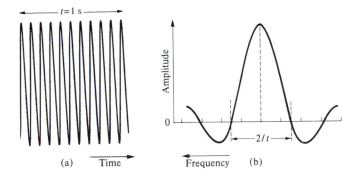

Fig. 5.13 (a) A pulse of radiation, and (b) its frequency distribution (only the central components of which are shown). This frequency distribution differs slightly from the frequency *response* to the pulse of an NMR system.

distribution Δv to be much greater than 5 kHz. Therefore, using $\Delta v \approx 1/t$, the duration of the pulse must be less than 200 μs.[1] Thus in Fourier-transform NMR the radiofrequency field is applied in the form of short pulses of sufficient power to ensure that during this short time the magnetization is tilted through a reasonably large angle.

The response of the nuclei to the pulse of radiofrequency field B_1 is totally analogous to the ringing of the tuning forks discussed above. Following the application of the pulse, an FID is generated. If all of the nuclei resonate at the same frequency, then the FID will appear as in Fig. 5.11(a), and Fourier transformation will give a single signal, as shown in Fig. 5.11(b). However, if there are several frequency components, reflecting a range of differing chemical environments, the resulting FID will be more complex; Fig. 5.12(a), which shows the ringing of three tuning forks, could equally represent the FID from nuclei in three different chemical environments. The Fourier transform of this decay would then give a conventional NMR spectrum as shown in Fig. 5.12(b).

5.7.2 The Fourier transform in MRI

If we consider a one-dimensional imaging experiment as outlined in Section 1.5.2, then the statements made above about the Fourier transform in spectroscopy apply equally to this simple imaging experiment, except that now the

[1] In fact, although this argument holds qualitatively, in quantitative terms the precise frequency *response* of the NMR system differs slightly from the frequency distribution of the *applied* pulse. The effects of finite pulse width are discussed in more detail in Section 7.3.5. Note also that if we wish to apply a selective pulse that excites signals from only a narrow frequency band, we can do this by using a long weak pulse of the appropriate frequency, for a long pulse has only a small frequency distribution.

frequency encodes for spatial information rather than for chemistry. However, imaging studies are rarely so straightforward, for at the very least a two-dimensional cross-sectional image is generally required. This, in turn, requires the use of two-dimensional mathematical reconstruction methods for generating the images from raw data.

Historically, the first method of NMR image reconstruction (Lauterbur 1973) was based not on Fourier analysis, but on a method known as projection-reconstruction, which had similarities to the approaches used in X-ray computed tomography. Then, in 1975, Kumar *et al.* proposed the use of two-dimensional Fourier transformation, following the successful implementation of such methods for the increasingly complex spectroscopic analyses of molecular structures in solution. This method, in a modified form that was termed spin-warp imaging (Edelstein *et al.* 1980), now forms a general basis for the majority of imaging studies. This is the approach that we now discuss.

A standard pulse sequence for two-dimensional Fourier transform (2D FT) imaging is shown in Fig. 5.14. The main features for consideration here are:

(1) the use of a frequency selective pulse (the 'shaped' 90° pulse in Fig. 5.14) to excite a single slice from within a three-dimensional region;

(2) the use of phase-encoding gradients, G_y, to provide information along one direction (the y-direction) within the slice; and

(3) the use of the read gradient, G_x, to provide information along the x-direction within the slice.[1]

We shall first discuss slice selection, before considering how the combined use of read and phase-encoding gradients can, with the aid of two-dimensional Fourier transformation, generate a cross-sectional image.

5.7.2.1 *Slice selection*

In the presence of a field gradient G_z, the field experienced by nuclei positioned at any location z is

$$B_{eff} = B_0 + G_z z \qquad (5.20)$$

Therefore, their resonance frequency is

$$\nu_{eff} = \frac{\gamma}{2\pi} (B_0 + G_z z) \qquad (5.21)$$

i.e. the resonance frequencies of the nuclei are determined by their positions along the z-axis. A frequency-selective radiofrequency pulse will therefore excite only a selected slice of nuclei, as shown in Fig. 5.15. The thickness of the slice is determined by the magnitude of the field gradient and the frequency

---

[1] For simplicity, we use the (arbitrary) coordinate convention that G_x is the read gradient, G_y is the phase-encoding gradient, and G_z is the gradient used for slice selection.

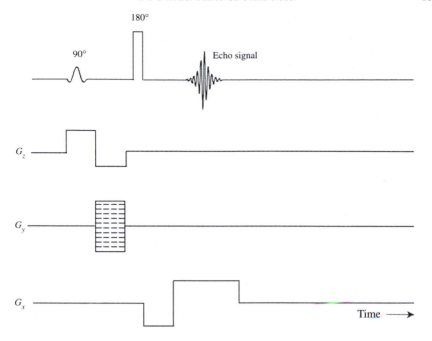

Fig. 5.14 Radiofrequency pulse and gradient timings for a 2D FT imaging sequence. G_z is the gradient used for slice selection, G_y is the phase-encoding gradient, and G_x is the read gradient. The dashed lines associated with G_y specify the fact that consecutive scans are acquired using a range of different G_y values.

bandwidth of the radiofrequency pulse; if the frequency bandwidth is $\Delta\nu$, it may be seen from eqn (5.21) that the slice thickness Δz will be given by

$$\Delta z = \frac{2\pi\Delta\nu}{\gamma G_z}$$

(5.22)

On the basis of the simple arguments presented in the above section, this bandwidth can be expected to be inversely proportional to the duration of the pulse: the longer the pulse, the narrower the bandwidth.

In general, it will be desirable to produce uniform excitation of the spins across the thickness of the selected slice, and to generate no excitation outside this slice. Radiofrequency pulses with a rectangular intensity profile have limitations in this respect, because their frequency response has a shape that is somewhat similar to the sinc, or $(\sin x)/x$ function that is shown in Fig. 5.13(b). As a result, the frequency selectivity is poor. Fourier analysis suggests that a much better approach would be to use pulses with a sinc profile, as these might

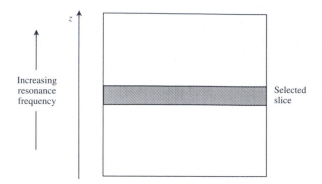

Fig. 5.15 In the presence of a field gradient G_z, the resonance frequency varies with position along the z-axis. Therefore, a frequency-selective radiofrequency pulse can be used to excite a selected slice of nuclei, as shown by the hatched region.

be expected to produce a rectangular frequency response.[1] In practice, while such pulses are now commonly used in both imaging and spectroscopy, further analysis has led to additional improvements in selectivity.

It may be seen from Fig. 5.14 that following the slice-selective 90° pulse the z-gradient is reversed. The reason for this is that the presence of the z-gradient during the pulse causes a variation of phase across the thickness of the slice, and if this were not compensated for there would be a resulting loss of signal. The gradient reversal following the pulse achieves this compensation, ensuring that the signals throughout the slice are in phase with each other.

5.7.2.2 *The read gradient in MRI*

Let us suppose that we wish to image the selected slice. As discussed in Section 1.5.2, the application of the read gradient G_x during the acquisition of data encodes for spatial location in the x-direction, and one-dimensional Fourier transformation would produce a projection as was shown in Fig. 1.7. This is exactly analogous to one-dimensional Fourier transformation in spectroscopy, the difference being that in spectroscopy the frequency distribution of the signals arises as a result of the different chemical environments of the detected nuclei, whereas in imaging it arises as a result of their differing spatial locations.

In practice, the signal in imaging studies is commonly collected not as a simple free induction decay, but as an echo that grows to its maximum value and then decays. The reasons for this, and for the fact that the gradient G_x contains a negative as well as a positive lobe, are discussed in Section 8.8.

---

[1] This argument reflects the fact that the sinc and rectangular functions form a Fourier pair; Fourier transformation of each function produces the other. However, as noted in Section 5.7.1, some caution with this type of analysis is necessary, as in quantitative terms the precise frequency *response* of the NMR system differs slightly from the frequency distribution of the applied pulse.

5.7.3.2 *Phase encoding in MRI*

In order to encode for the y-direction too, it is necessary to make use of y-gradients. In the 2D FT method shown in Fig. 5.14, these are applied prior to data acquisition in a series of consecutive scans, the effects of which we now consider.

In the presence of a field gradient G_y, the field experienced by nuclei at any location y is

$$B_{\text{eff}} = B_0 + G_y y \qquad (5.23)$$

Therefore, their resonance frequency is

$$\nu_{\text{eff}} = \frac{\gamma}{2\pi}(B_0 + G_y y) \qquad (5.24)$$

and the frequency offset caused by the gradient is

$$\Delta\nu = \frac{\gamma}{2\pi}G_y y \qquad (5.25)$$

As a result, the nuclei precess at different rates according to their y-position, and after a time t undergo a phase shift

$$\Delta\phi = 2\pi\Delta\nu t$$
$$= \gamma G_y y t \qquad (5.26)$$

It may be seen that the gradient G_y causes the nuclei to acquire a phase shift that is proportional to their y-coordinate, the gradient strength, and the time for which the gradient is applied. This phase shift is 'remembered' by the nuclei during the sequence shown in Fig. 5.14, and generates a phase shift in the ensuing signal. We shall now consider the nature of these signals.

When the gradient $G_y = 0$, no phase shifts at all are obtained, and the signals from all y-coordinates are equivalent. Representative signals for nuclei at three representative y-positions ($y = 0$, $y = Y/2$, and $y = Y$) are given in Fig. 5.16(a). Let us now, in a second scan, apply a gradient G_{y180} whose strength and duration is such that at the outermost regions of the slice ($\pm Y$) the resulting phase change is 180°. From eqn (5.26), we see that the signal from nuclei at $y = 0$ remains unchanged, while that from nuclei at $y = Y/2$ changes phase by 90°. The signal from nuclei at $y = Y$ is inverted as shown in Fig. 5.16(b). In subsequent scans, there is a stepwise increase in gradient strength such that at $y = Y$ there is a stepwise phase change of 180°. The effect of this is that, as the gradient strength increases, the signal from $y = 0$ remains unchanged, the signal from $y = Y/2$ undergoes successive 90° phase shifts such that at every fourth scan the signal repeats, while the signal from $y = Y$ repeats every second scan. In other words, the stepwise increases in the phase-encoding gradient cause the signals to vary or modulate with a frequency that is determined by the y-coordinate of the nuclei generating the signal (see

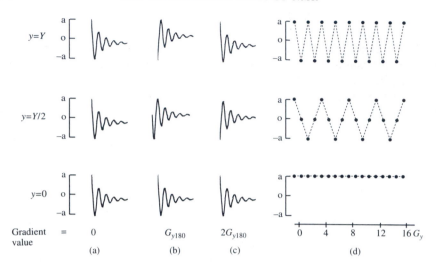

Fig. 5.16 The principles of phase encoding. (a) When there is no field gradient, the nuclei at the three y-coordinates all give rise to a signal of the same phase. (b) In the presence of a field gradient G_{y180}, the phase of the nuclei at $y = 0$ remains unchanged, while the phases of the nuclei at $y = Y$ and at $y = Y/2$ change by 180° and 90° respectively. (c) When the gradient strength is doubled, the phases of the nuclei at these two coordinates change by a further 180° and 90° respectively. (d) As the gradient G_y is stepped through a range of values, the initial amplitudes of the signals change as shown. The frequencies at which these amplitudes change are directly related to the y-coordinates of the nuclei.

Fig. 5.16(d)). The actual signal that is stored within the computer is a complex summation of the signals generated by all of the nuclei. Fourier analysis of how the signals vary from one scan to another provides a means of disentangling the frequency modulations and hence of yielding the y-distribution of the nuclei that give rise to the signals. This provides one dimension of the Fourier analysis. The second is provided by Fourier analysis of the FID signal variations during the application of the read gradient G_x. The combination of the two, using two-dimensional Fourier transformation, produces full two-dimensional image reconstruction.

In practice, the number of phase-encoding steps is typically 128 or 256, in order to generate the required spatial resolution along the y-axis. Data acquisition may begin with the strongest positive y-gradient, which is then gradually stepped through zero to the strongest negative y-gradient. However, while it may be convenient and logical to step through the gradients in this way, this is not strictly necessary; one could, for example, alternate between positive and negative gradients. Such alternative procedures can be described mathematically in terms of alternative trajectories through so-called k-space, and we finish the chapter with a brief discussion of what is meant by k-space.

5.8. THE CONCEPT OF *k*-SPACE

While the signals that emerge from the receiver are continuous, the computer samples the incoming signals at discrete time intervals (see Section 7.6). For example, in imaging studies the free induction decay may be sampled, say, 256 times in the presence of the read gradient. If there are also 256-phase encoding steps, then the acquired data set can be considered as a two-dimensional (256 × 256) array of points that are acquired in what is termed *k*-space. Fourier transformation of this data set will generate the same number of data points in the resulting two-dimensional image.

In order to understand what is meant by *k*-space, it is convenient to return to a simple spectroscopy study, in which a free induction decay is acquired in the form of an oscillating signal varying as a function of time. We have seen that an FID of the form shown in Fig. 5.11(a) will give rise to a spectrum consisting of a single component of a fixed frequency (Fig. 5.11(b)). An FID that oscillates at a lower frequency would, of course, give rise to a spectrum consisting of a single line at the corresponding lower frequency. Conversely, inverse Fourier transformation of the signal in Fig. 5.11(b) would give again the corresponding FID. This inverse process reflects the reciprocal relationships between the frequency and time domains. As a further reflection of this reciprocal relationship, we could interchange the frequency and time domains, so that we could consider the oscillation of Fig. 5.11(a) to be a spectrum rather than an FID. Of course, it would be a somewhat unusual spectrum, and the corresponding FID would be equally unusual, for the Fourier-transformation process shows that the FID would look just like Fig. 5.11(b), i.e. it would be zero at almost all time points. Clearly, this is not a very realistic FID, but it does illustrate what would happen if one point in the FID were artefactually high (in fact Fig. 5.10 illustrates this more clearly); it would generate an oscillating baseline in a spectrum, the frequency of such an oscillation increasing for points further away from zero time. This argument becomes more realistic if we consider a simple one-dimensional imaging study, in which the image is generated by Fourier transformation of an FID acquired in the presence of a read gradient. Suppose we have a structure in which the proton density has a cyclical distribution; this is well represented by Fig. 5.10(a). By direct analogy with the above argument, the FID would consist of a single point, as in Fig. 5.10(b). If the structure had a lower spatial frequency, the FID would again consist of a single point, but it would be closer to zero time. Thus different points in the FID encode for different spatial frequencies, points closer to zero time encoding for lower spatial frequencies. In practice, this means that data points towards the end of the free induction decay provide the information required for fine spatial resolution, and hence contribute detail to the resulting image. Conversely, data points close to zero time in the free induction decay contribute to the relatively coarse overall shape of the image.

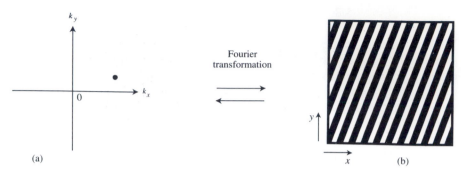

Fig. 5.17 A two-dimensional structure of the form shown in (b) can be represented by a single data point in two-dimensional k-space (a).

If we say that data are collected in k-space, then it may be seen from the above that, when considering the properties of a free induction decay, k-space has close analogies with time; points close to zero time correspond to small spatial frequencies, i.e. small values of k, whereas points further away from zero time correspond to higher spatial frequencies, i.e. higher k-values. The concept of k-space takes on additional value when it is extended to two, three, or more dimensions. In particular, if we consider the phase-encoding direction, analogous arguments show that the data obtained in the presence of the highest phase-encoding gradients (i.e. those that give rise to the highest phase shifts) correspond to high values of k, and add fine detail in the phase-encoding direction, while data acquired with the low gradients correspond to lower values of k, and give shape to the images. We then have a two-dimensional k-space, characterized by k_x, representing the read direction, and k_y representing the phase-encoding direction.

To take a simple example of this, Fig. 5.17(b) shows a grid structure which can be specified by a single relatively high spatial frequency in the x-direction, and a single relatively low spatial frequency in the y-direction. Suppose that we were to image this structure using a standard two-dimensional sequence as illustrated in Fig. 5.14. On the basis of the above argument, we might expect the raw imaging data set to be represented by a single data point[1] corresponding to the respective spatial frequencies, as shown diagramatically in Fig. 5.17(a). This example serves to illustrate what happens when a single data point in an imaging data set is artefacted; it generates a series of stripes in the resulting image, the nature of the stripes depending on the particular location of that data point in k-space.

A more typical example is shown in Fig. 5.18, which shows the raw imaging

---

[1] This is an approximation, given the fact that the structure is finite and has sharp boundaries.

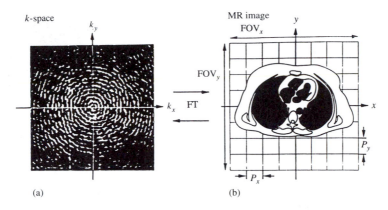

Fig. 5.18 The raw data in a two-dimensional Fourier transform imaging sequence can be displayed in the form of a grid in k-space (a). Fourier transformation gives rise to the image shown in (b). FOV is field-of-view, and P is pixel size (Reprinted with permission from Stehling *et al.* (1991), *Science*, **254**, 43–50. Copyright 1991 American Association for the Advancement of Science).

data that correspond to a more realistic object. The imaging data are represented as a rectangular grid, i.e. as a series of lines displayed one above the other. The central line represents, point by point, the observed signal for $k_y = 0$, (i.e. the data obtained when the phase-encoding gradient $G_y = 0$), while each of the other lines represents data obtained at each of the different non-zero values of G_y. In each line, the brightness of each point represents the amplitude of the signal, and so the oscillating (or fluctuating) signal is no longer represented as, for example, in Fig. 5.11(a), but as an oscillation in brightness. The two data sets in Fig. 5.18 can be interconverted by two-dimensional Fourier transformation.

One of the questions that arises is whether there are optimal procedures for collecting data in the spatial frequency or k-space domain. There is little scope for changing the manner in which data are collected during the read gradient, for essentially there is little alternative to collecting the data simply as a function of time. This is equivalent to saying that in spectrosopy the free induction decay has to be collected as a function of time; it is simply not possible to collect a later time point in the FID prior to an earlier time point. However, there is scope for manipulating the phase encoding gradients so that, for example, one does not simply proceed stepwise from the highest positive gradient through zero to the highest negative gradient. Indeed, there are many ways of collecting two-dimensional data sets that can be transformed into two-dimensional images, and each of these can be represented in terms of different trajectories through k-space. For example, echo-planar imaging, which enables a complete two-dimensional data set to be acquired in, say, 40 ms, involves

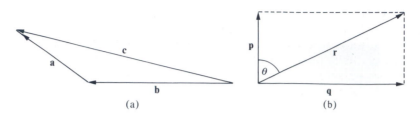

Fig. 5.19 (a) Vector diagram in which *c* is the resultant of *a* and *b*. (b) Vector diagram in which *r* is resolved into two perpendicular components *p* and *q*.

a different trajectory through *k*-space from standard two-dimensional Fourier imaging techniques (see Section 8.8). Analysis of *k*-space provides a means of analysing these different imaging techniques, of evaluating their relative advantages and disadvantages, and of understanding (and thereby reducing or eliminating) the degradative effects of artefactual data points or of motion.

APPENDIX

In this appendix, we introduce some of the mathematics underlying the concepts that have been discussed in the chapter. We begin with some comments about vectors, before proceeding to a discussion of complex numbers, which leads finally to some of the equations involved in Fourier analysis.

A1. Vectors

Scalar quantities such as temperature, mass, and length are specified by magnitude alone. However, some physical quantities are only completely specified if their direction is given in addition to their magnitude. Such quantities are known as vectors. Magnetic field, angular momentum, and magnetic moment are all vector quantities; angular momentum, together with other quantities that describe rotational motion, is specified by a vector, the direction of which gives the axis about which rotation takes place (see under vector multiplication below). Vector algebra therefore figures prominently in basic NMR theory.

A1.1 *Vector addition and subtraction*

A vector can be represented diagrammatically by an arrow pointing along the direction of the vector. The length of the arrow reflects the magnitude of the vector quantity. The addition of two vectors is performed using the triangle law of addition as illustrated in Fig. 5.19(a). The vector *c* is the sum or resultant of the two vectors *a* and *b*. The difference *a* − *b* is found in a similar manner, except that the direction of *b* is reversed.

Any vector *r* can be resolved into two components at right angles; this is effectively the converse of vector addition. Figure 5.19(b) shows how r is resolved into two perpendicular components p and q. The component of r along the direction of p is given by the magnitude of p, and is equal to $r \cos \theta$. Similarly, the component of r along q is $r \sin \theta$. Any vector r can be resolved in a similar manner into three components parallel to the x-, y-, and z-axes of a Cartesian coordinate system.

A1.2 Vector multiplication

Multiplication of a vector quantity by a scalar quantity, s, simply involves multiplying the magnitude of the vector by the factor s and has no effect on the direction of the vector. Thus, r multiplied by s is equal to sr.

It is often convenient to introduce the unit vectors i, j, and k as vectors of unit magnitude that are parallel to the x-, y-, and z-axes respectively of a Cartesian coordinate system. A vector r can then be written

$$r = ai + bj + ck$$

where a, b, and c are the magnitudes of the components of r parallel to the three axes.

The *scalar product* of two vectors p and q is written $p \cdot q$, and is sometimes called the *dot product*. It is a scalar quantity of magnitude $pq \cos \theta$, where θ is the angle between p and q. The scalar product of two perpendicular vectors is zero, since $\cos 90° = 0$. Thus $i \cdot i = j \cdot j = k \cdot k = 1$, whereas $i \cdot j = j \cdot k = k \cdot i = 0$.

An example of the use of a scalar product is given by the interaction of the magnetic moment μ of a nucleus with an applied field B_0. The energy of this interaction is $E = -\mu \cdot B_0$, which leads to the result $E = -\gamma \hbar m B_0$ given in eqn (5.6). Perhaps a more familiar example is the work, W, done in moving an object a distance ds against a force F: $W = -F \cdot ds$.

An alternative form of multiplication of two vectors is known as the *vector product* or *cross product*. The vector product of p and q is a vector of magnitude $pq \sin \theta$ and direction perpendicular to both p and q. The vector product is written $p \times q$, and its direction is that in which a right-handed screw would move if turned from p to q. Thus $p \times q = -q \times p$, since the first vector product is equal to but opposite in direction to the second. Note that $i \times i = j \times j = k \times k = 0$ and that $i \times j = -j \times i = k$.

An example of the vector product is given by the expression for the angular momentum p of an object about a point, O. If v is the velocity of the object, m is its mass, and r is the position vector of the object relative to O, then the angular momentum p is given by

$$p = r \times mv$$

It can be seen from Fig. 5.20 that the direction of p is along the axis of rotation.

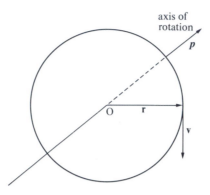

Fig. 5.20 Diagram illustrating that the direction of the angular momentum vector p is along the axis of rotation.

Another example of the vector product is given by the equation of motion of a magnetic moment μ in a field B_0. The equation of motion is

$$\frac{d\mu}{dt} = \gamma\mu \times B_0$$

The motion represented by this equation is the precession of the magnetic moment about B_0 with an angular frequency γB_0; thus the equation describes Larmor precession.

A1.2 Complex numbers

Consider the quadratic equation $x^2 - x - 1 = 0$. This can easily be solved to give $x = \{1 \pm \surd(-3)\}/2$, i.e. $x = \frac{1}{2} \pm i\surd3/2$ where $i = \surd(-1)$. The two solutions to this equation are complex numbers, comprising a real part ($\frac{1}{2}$) and an imaginary part ($\pm i\surd3/2$). The term imaginary implies that such numbers have no real physical meaning; nevertheless they play an extremely useful and elegant role in the analysis of waveforms. Complex numbers are therefore of considerable value in the theory of Fourier-transform NMR, and it is primarily for this reason that this section is included.

Complex numbers are often represented by an Argand diagram in which the real part is represented by the x-coordinate, and the imaginary part by the y-coordinate, as shown in Fig. 5.21. The number $a + ib$ is therefore totally specified by the vector $\overrightarrow{OP}$.

Let us consider multiplication of this number by i. The product is $ia + i^2b$, which is equal to $ia - b$ since $i^2 = -1$. This is represented by the vector $\overrightarrow{OR}$. Note that the vector $\overrightarrow{OR}$ has the same magnitude as $\overrightarrow{OP}$, but has been rotated through 90°. Multiplication by i corresponds to a rotation, or phase shift, of 90°.

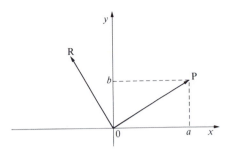

Fig. 5.21 An Argand diagram.

This interpretation of multiplication by i is not confined simply to the Argand diagram representation; if we consider the number $e^{i\phi}$, which can alternatively be written $\exp(i\phi)$, it may be shown that

$$\exp(i\phi) = \cos\phi + i\sin\phi$$

Expressions of the form $A\exp(i\phi)$ are often used to describe waveforms because they contain information about amplitude (A) and about phase (ϕ). Moreover, such expressions are often far more amenable to mathematical analysis than cosine and sine terms. If we consider multiplication by i, it may readily be shown that

$$i\exp(i\phi) = \exp\{i(\phi + \pi/2)\}$$

i.e. multiplication by i is equivalent to a shift in phase by $\pi/2$ radians or $90°$, in agreement with the Argand diagram representation.

A1.3 The Fourier transform in NMR

There is no unique definition of the Fourier transform, but a convenient formal definition is

$$F(\omega) = \int_{-\infty}^{\infty} f(t) \exp(i\omega t)\, dt$$

The inverse transform is

$$f(t) = \frac{1}{2\pi} \int_{-\infty}^{\infty} F(\omega) \exp(-i\omega t)\, d\omega$$

These expressions enable a time-dependent function $f(t)$ to be analysed in terms of its frequency components $F(\omega)$ and vice versa.

The response of the nuclear spins to a radiofrequency pulse can be expressed by

$$f(t) = A\cos\omega_0 t \exp(-t/T_2^*)$$

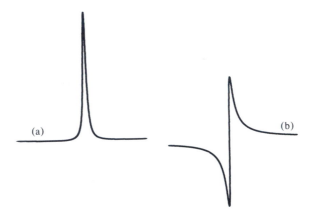

Fig. 5.22 The Fourier transform of the FID shown in Fig. 5.8(a). (a) is the real component, and (b) is the imaginary component.

where ω_0 is the Larmor frequency and $\exp(-t/T_2^*)$ expresses the fact that the magnetization decays with a time constant T_2^*. The Fourier transform of this response is

$$F(\omega) = \int_0^\infty A \cos \omega_0 t \exp(-t/T_2^*) \exp(i\omega t)\,dt$$

where the lower limit in the integral is zero because the time response is zero prior to the application of the pulse. On solving this integral we obtain a real component $F_{\text{real}}(\omega)$ given by

$$F_{\text{real}}(\omega) = \frac{A}{2} \frac{T_2^*}{1 + (T_2^*)^2(\omega - \omega_0)^2}$$

and an imaginary component

$$F_{\text{imag}}(\omega) = \frac{iA}{2} \frac{\omega - \omega_0}{1 + (T_2^*)^2(\omega - \omega_0)^2}$$

The form of these two functions is shown in Fig. 5.22. The real part of the Fourier transform is precisely the Lorentzian lineshape given by eqn (1.8), and therefore in this example is equivalent to the absorption shape obtained in continuous-wave NMR. However, what about the second, imaginary term? We have seen above that the imaginary term can be regarded as representing a component 90° out of phase with the real component. In NMR it can be shown that the two terms correspond to the components of magnetization along two perpendicular axes of the rotating frame. Thus, if the real component represents $M_{x'}$, the imaginary component represents $M_{y'}$. It turns out

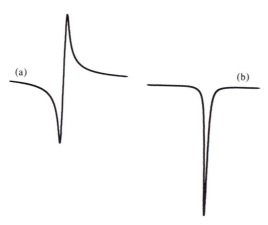

(a)

(b)

Fig. 5.23 The Fourier transform of the FID shown in Fig. 5.8(b). (a) is the real component and (b) is the imaginary component.

that these two terms are interdependent (one specifies the other) and they have as much physical significance as each other. In the particular case we have considered, the real part corresponds to the absorption mode and the imaginary part to what is called the dispersion mode (see below). However, if we had considered the transform of the decay $A \sin \omega_0 t \exp(-t/T_2^*)$, we would have found that the real part would now contain the dispersion mode while the imaginary part would contain a negative absorption signal (see Fig. 5.23). This corresponds to a 90° phase shift. More generally, the free induction decay will be represented by a function of the form $A \cos(\omega_0 t + \phi)$, the phase angle ϕ being somewhat arbitrary because it will depend upon the way in which the NMR system is set up. Therefore, in practice, the real and imaginary parts obtained on Fourier transformation will each contain a mixture of the absorption and dispersion modes, and a phase correction will be required in order to obtain the required absorption and dispersion modes (see Section 7.6.2).

The dispersion mode is given its name for the following reason. In the optical region of the electromagnetic spectrum, variation of the refractive index with wavelength or frequency has been known about for a long time and is called dispersion. Usually the refractive index increases with frequency and this is known as 'normal dispersion'. However, in the vicinity of an absorption line the reverse happens; the refractive index decreases with increasing frequency, a phenomenon known as 'anomalous dispersion'. In a simple case the absorption and dispersion are given by the two terms $F_{real}(\omega)$ and $F_{imag}(\omega)$, respectively, that are given above; hence the terminology that is used.

REFERENCES

Edelstein, W. A., Hutchison, J. M. S., Johnson, G., and Redpath, T. (1980). Spin warp NMR imaging and applications to whole body imaging. *Phys. Med. Biol.*, **25**, 751–6.

Ernst, R. R. and Anderson, W. A. (1966). Application of Fourier transform spectroscopy to magnetic resonance. *Rev. Sci. Instrum.*, **37**, 93–102.

Kumar, A., Welti, D., and Ernst, R. R. (1975). NMR Fourier zeugmatography. *J. Magn. Reson.*, **18**, 69–83.

Lauterbur, P. C. (1973). Image formation by induced local interactions: examples employing nuclear magnetic resonance. *Nature*, **242**, 190–1.

Roth, K. (1984). *NMR-tomographie und -specktroscopie in der medizin*. Springer Verlag, Berlin.

Stehling, M. K., Turner, R., and Mansfield, P. (1991). Echo-planar imaging: magnetic resonance imaging in a fraction of a second. *Science*, **254**, 43–50.

6

The NMR parameters

Detailed interpretation of magnetic resonance images and spectra requires an understanding of the various factors that influence the characteristics of NMR signals. In previous chapters, it was noted that the nature of NMR signals depends on a wide range of physical and chemical effects, and the purpose of this chapter is to explain the basis of these effects, and to indicate how they can be quantified. We shall consider in turn the chemical shift, spin–spin coupling, and relaxation, and then discuss the ways in which these properties of NMR signals can be influenced by magnetization transfer effects and by paramagnetic species.

6.1 THE CHEMICAL SHIFT

Of the parameters that characterize NMR signals, the chemical shift was the one that most opened up the technique to the scientific community. The existence of the chemical shift enables us to use NMR to distinguish not only between different molecules, but also between individual atoms within a molecule. When used in conjunction with intensity measurements and spin–spin coupling data, the chemical shifts of the spectral lines of a molecule provide a great deal of information about its structure. NMR is therefore an essential structural tool for organic chemists, and it provides an invaluable aid to structure determinations of biological molecules (see, for example, Roberts 1993).

In Section 1.5 a brief introduction was given to the concept of the chemical shift. In particular, it was noted that an applied field B_0 induces electronic currents in atoms and molecules, and that these produce an additional small field at the nucleus proportional to B_0. The total effective field at the nucleus can therefore be written

$$B_{\mathrm{eff}} = B_0 - B_{0\sigma}$$
$$= B_0(1 - \sigma) \qquad (6.1)$$

σ is called the shielding or screening constant because the effect of these electronic currents is to shield the nuclei from the effects of the applied field. σ is sensitive to the chemical environment of the nuclei, and therefore nuclei in different chemical environments experience different fields and hence produce signals at different frequencies. The separation of resonance frequencies from an arbitrarily chosen reference frequency is termed the chemical shift. The secondary shielding fields generated by the electrons are very small in comparison with the applied field, and so the absolute spread in frequency is also small. As a result, it can often be difficult to resolve the various resonances, particularly those of large molecules that produce complex spectra. Since the frequency separation of resonances is proportional to B_0, an increase in B_0 will tend to spread them out and hence improve the spectral resolution.

In this section we discuss some of the factors that contribute to chemical shifts and describe the practicalities and conventions involved in chemical shift measurements.

6.1.1 Conventions and terminology

1. Chemical shifts are expressed in terms of the dimensionless unit of parts per million (ppm), and as such their values are independent of the magnitude of the field B_0. The chemical shift δ is defined as

$$\delta = \frac{\nu_S - \nu_R}{\nu_R} \times 10^6 \qquad (6.2)$$

where ν_S is the resonance frequency of the nuclei of interest and ν_R is the frequency of an arbitrarily chosen reference. Using eqn (6.1), we find that

$$\delta = \frac{\sigma_R - \sigma_S}{1 - \sigma_R} \times 10^6$$

i.e.

$$\delta = (\sigma_R - \sigma_S) \times 10^6 \qquad (6.3)$$

since $\sigma \ll 1$. Therefore the chemical shift is obtained simply from the difference between the shielding constants of the nuclei of interest and of the reference nuclei. Equation (6.3) expresses the accepted convention that the chemical shift of a resonance is negative if the nucleus is more shielded than the reference.

2. The term 'upfield' and 'downfield' are sometimes used to denote the direction of a chemical shift. These terms originated in the days of continuous-wave NMR when spectra were obtained by using a fixed-frequency source and sweeping through the applied field B_0, and it is useful to consider their derivation. Suppose that a weakly shielded nucleus resonates at a fixed frequency ν_0 in an applied field B_0. A nucleus that is more strongly shielded experiences a lower effective field, and therefore, to bring it to resonance at the frequency ν_0, the applied field has to be increased. Its resonance is therefore said to be shifted upfield. If we now consider Fourier-transform NMR, a fixed field is applied and resonances are separated according to their frequency. Again, the weakly shielded nucleus resonates at a frequency ν_0 in

the field B_0. However, the more strongly shielded nucleus experiences a reduced field and must therefore resonate at a lower frequency. Thus an increase in shielding produces an *upfield* shift if the field is varied, but a shift to *low* frequency when frequency is the variable. Intuitively, one might have expected that, since frequency is proportional to field, upfield would have corresponded to high frequency; in fact, it corresponds to low frequency. Although the advent of Fourier-transform NMR has made the terminology of upfield and downfield shifts obsolete and confusing, these terms are still in common usage and therefore have to be tolerated for the present.

3. Spectra are plotted according to the traditional rule of spectroscopy: increasing wavelength, or decreasing frequency, to the right.

To summarize, if one nucleus is more shielded than another, its signal will be shifted to low frequency (or equivalently upfield); by convention it will have a more negative chemical shift, and will appear further towards the right-hand side of the spectrum.

6.1.2 Internal and external standards and the effects of magnetic susceptibility

The local field experienced by a nucleus is modified not only by the shielding effects of the electrons but also by shielding produced by the surrounding medium. This 'bulk shielding' modifies the field by an amount B_{bs}

$$B_{bs} = S_f \chi B_0$$

where χ is the magnetic susceptibility of the medium and S_f is a numerical factor that depends on the orientation and shape of the sample. The magnitude of χ is of the order of 10^{-5} for common materials, and therefore the bulk shielding cannot in principle be ignored. Magnetic susceptibility is discussed in more detail in Section 6.5 in relation to the effects of para-magnetic species.

Chemical shifts are measured by comparing an observed resonance frequency with that of a reference compound. When the reference is added to or is intrinsic to the sample, it is termed an *internal* reference. Under these conditions the reference and sample both experience the same bulk shielding and no corrections for susceptibility effects are necessary; observed chemical shift differences must be due entirely to local interactions. However, if it is not feasible to add the reference directly to the sample, an *external* reference may be required. Under these circumstances it may be necessary to correct the observed shift for the effects of susceptibility differences between the reference and sample. The problem is that the shift associated with these effects depends not only on the susceptibility difference but also on the orientation and shape of the sample and reference. For these reasons, care needs to be taken with the practical use of external standards (such as 85 per cent phosphoric

Fig. 6.1 The orbital motion of electrons, induced by the field B_0.

acid in the case of ^{31}P NMR). Clearly, it is preferable, whenever possible, to use an internal reference. For example, the ^{31}P signal of phosphocreatine in skeletal and cardiac muscle, and in brain, can provide a convenient and acceptable internal reference (see Section 2.7); in fact, many ^{31}P studies of tissue metabolism assign a chemical shift of 0 ppm to the phosphocreatine signal. Others continue to use the more traditional chemical shift scale on which the phosphoric acid signal would appear at 0 ppm (see, for example, Fig. 2.9).

6.1.3 Contributions to chemical shifts

A number of factors can contribute towards chemical shifts. We can distinguish between local and long-range contributions.

6.1.3.1 *Local diamagnetic effects*

Local diamagnetic effects, which we considered briefly above, exist for all atoms and molecules, and arise because an applied field B_0 induces orbital motion in the electrons, as shown in Fig. 6.1. This motion generates a secondary magnetic field at the nucleus that opposes B_0 and therefore 'shields' the nucleus from the applied field. The magnitude of the local diamagnetic shielding is determined primarily by the local electron density. Therefore, if this is the dominant shielding mechanism, we should expect factors such as the electronegativity of adjacent atoms to be reflected in the magnitude of the observed shifts.

For protons, the local shielding is often dominated by the diamagnetic term, but for nuclei such as ^{13}C and ^{19}F, additional more complicated effects become important. These so-called paramagnetic effects (an explanation of which is beyond the scope of this book) enhance the applied field B_0, in contrast to the diamagnetic contribution that opposes B_0.

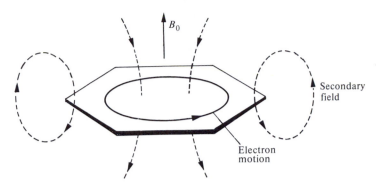

Fig. 6.2 Ring currents, and the secondary fields that they generate.

6.1.3.2 *Long-range shielding*

The local effects considered above arise from shielding by electrons on the same atom as the nucleus in question. Shifts can also arise from long-range shielding caused by the induced motion of electrons associated with *other* atoms or groups of atoms. It can be shown that such long-range shielding occurs only if the motion of the neighbouring electrons is anisotropic, i.e. if the motion differs according to the orientation of the molecule in the applied field.

Ring-current shifts provide a well-known example of such effects. In aromatic systems the delocalized π electrons give rise to a ring current if the applied field B_0 has a component perpendicular to the plane of the ring (see Fig. 6.2). This current generates a secondary field which has a direction and magnitude at a given nucleus that is highly dependent on the position of the nucleus relative to the plane of the ring. For example, an aromatic proton in the plane of the ring experiences an enhanced field, whereas a proton above or below the ring experiences a reduced field. Ring-current shifts can be large (of the order of 1 ppm) and extend over considerable distances, typically up to about 6 Å.

Shifts arising from paramagnetic centres provide a second example of long-range shielding, as discussed in Section 6.5.

6.1.4 Chemical shift anisotropy

If a group within a molecule is anisotropic, the extent of shielding at a nucleus near or within the group will depend upon the orientation of the group relative to the applied field. We have noted this effect in the case of aromatic ring systems. There is no shielding if the plane of the ring is parallel to the applied field, but there are significant effects if the ring is perpendicular to the field. Thus the chemical shift of a nucleus near the ring will vary as the molecule containing the ring tumbles randomly in solution, and in fact will be averaged by the effects of motion to the value that is detected in the spectrum of the solution. This average, as we have seen, depends upon the precise position of the nucleus relative to the ring.

When the chemical shift of a nucleus depends on the orientation of the molecule within the applied field, the chemical shift is said to be anisotropic. If there is axial symmetry, the chemical shift anisotropy can be expressed in terms of the difference $\sigma_\parallel - \sigma_\perp$, where $\sigma_\parallel$ and $\sigma_\perp$ refer to the components of shielding parallel and perpendicular to the axis of symmetry of the grouping, respectively. Chemical shift anisotropy can make important contributions to the spectra of solids and highly immobilized systems such as membranes, because in such systems its effects are not averaged out by molecular motion.

Chemical shift anisotropy can also, as mentioned later in Section 6.3, provide a mechanism for spin–spin and spin–lattice relaxation.

6.2 SPIN–SPIN COUPLING

It was mentioned in Section 1.5 that resonances are often split into two or more components, as illustrated by the ^{31}P NMR spectrum of ATP shown in Fig. 1.6. This splitting arises from an interaction between neighbouring nuclear spins which is transmitted by means of the electrons in the bonds joining the nuclei. An explanation for the effect can be given most simply by considering an example of two nuclei, A and X, both of spin $\frac{1}{2}$, in atoms that are joined by a covalent bond.

The possible spin states of these two nuclei are illustrated in Fig. 6.3. In general, it can be shown that the electron-mediated interaction between the two nuclei is more favourable (i.e. of lower energy) when the nuclear spins are antiparallel (as in (i) and (ii) in the figure) than when they are parallel, as in (iii) and (iv). Now the resonance absorption of energy by nucleus A involves a change in its spin from $+\frac{1}{2}$ to $-\frac{1}{2}$, with no change in the spin of X. Therefore, it corresponds to a transition either from (ii) to (iii), or from (iv) to (i), as illustrated by the dotted lines in the figure. The amounts of energy absorbed in these two types of transition differ slightly from each other, because there is a net increase in the spin–spin interaction energy in one type, and a decrease in the other. Thus the resonance of nucleus A is split into two components, which are of equal intensities because the two types of transition are equally probable.

The separation of the two components is J_{AX} Hz, where J_{AX} is a constant known as the spin–spin coupling constant. The coupling constant J_{AX} is a characteristic of the molecule; it is independent of the magnitude of the applied field B_0, and this is why it is expressed in hertz rather than in parts per million. Since the interaction is mutual, the X resonance is also split into two components separated by J_{AX} Hz. Coupling constants have typical values in the region 1–200 Hz.

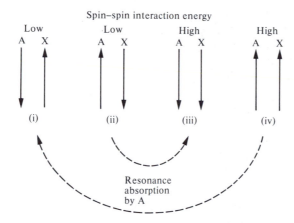

Fig. 6.3 This diagram illustrates the four possible spin states that exist for two coupled nuclei A and X, both of spin $\frac{1}{2}$. The $m = +\frac{1}{2}$ states and $m = -\frac{1}{2}$ states are represented by upward-pointing and downward-pointing arrows, respectively. The spin–spin interaction energy is higher in states (iii) and (iv) than in states (i) and (ii). Therefore less energy is absorbed in the transition from state (iv) to (i) than in the transition from (ii) to (iii).

Spin–spin coupling can occur when two nuclei are bonded together, e.g. ^{13}C–^{1}H, or when they are separated by more than one bond, e.g. ^{1}H–C–C–^{1}H. The coupling is transmitted through the intervening chemical bonds but tends to decrease as the number of bonds increases. Often a superscript is used, e.g. $^{3}J_{AX}$, to indicate the number of bonds between the coupled nuclei. When there are more than two magnetic nuclei in a molecule, coupling occurs between each pair of nuclei and the splitting pattern can become very complicated. Coupling patterns can be used empirically as a means of identifying chemical groupings within a sample. Often this identification relies upon double-resonance or spin-echo techniques which, as described in Section 8.11, can also be used for spectral editing, i.e. for the selective detection of specific signals of interest.

6.2.1 The analysis of spin–spin coupling patterns

The theoretical treatment of spin–spin coupling can only be considered in terms of quantum mechanics, and is outside the scope of this book. Here, therefore, we discuss the important rules governing spin–spin coupling and illustrate them by means of a few simple examples.

We begin by distinguishing between first-order and second-order spectra. First-order spectra are ones in which the frequency difference between the resonances of the nuclei involved in the spin–spin coupling is much greater than the magnitude of the coupling constant J. Second-order spectra, which are more difficult to interpret, arise when this condition is not totally fulfilled, i.e. when J is not small in comparison with the frequency difference.

There is an accepted nomenclature for spin systems whereby coupled nuclei that generate second-order spectra are assigned adjacent letters in the alphabet; otherwise they are assigned letters that are well separated. The letters conventionally used are A, B, and C; M and N; and X, Y, and Z. Thus an AX system produces first-order spectra, while an AB system generates second-order spectra.

6.2.2 First-order spectra

The rules governing first-order spectra are as follows.

1. The spin–spin coupling between the nuclei of an equivalent group, e.g. the three protons of the methyl group of acetaldehyde, produces no observable splitting. This is because of the existence of quantum mechanical selection rules that prohibit the appropriate transitions.

2. A nucleus coupled to n equivalent nuclei of spin I gives rise to $2nI + 1$ lines with relative intensities given by the binomial distribution. Thus, a nucleus coupled to a single nucleus of spin $\frac{1}{2}$ would produce a doublet with relative intensities 1:1; coupling to two equivalent nuclei of spin $\frac{1}{2}$ would produce a triplet with relative intensities 1:2:1; and coupling to three equivalent nuclei of spin $\frac{1}{2}$ would generate a quartet with relative intensities 1:3:3:1; etc.

3. If there is coupling to more than one group of interacting nuclei, a multiplet of multiplets is formed by straightforward extension of rule (ii).

4. The spacing within each multiplet due to a given interaction is equal to the coupling constant J for that interaction.

The ^{1}H spectrum of acetaldehyde shown in Fig. 6.4, and the ^{31}P spectrum of ATP that was shown in Fig. 1.6, provide simple examples of first-order spectra. In the ATP spectrum the signal from the γ-phosphate is split into two by interaction with the ^{31}P nucleus of the β group. The signal from the β-phosphate is split into a doublet of doublets by the α and γ ^{31}P nuclei. Since the coupling to both of these nuclei is the same, this results in the appearance of a triplet of lines of relative intensities 1:2:1. The signal from the α-phosphate is split into a doublet by the ^{31}P nucleus of the β group, and there is a smaller interaction with the CH_2 protons of the adenine moiety which would, if the resolution were sufficiently good, produce a subsidiary splitting of the two doublet lines.

6.2.3 Second-order spectra

Typical second-order spectra for two coupled nuclei A and B, both of spin $\frac{1}{2}$, are shown in Fig. 6.5. This figure illustrates what happens to the splittings

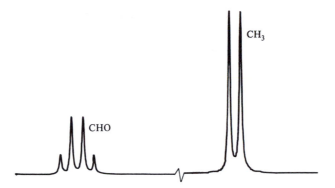

Fig. 6.4 ^{1}H NMR spectrum of acetaldehyde (CH_3CHO). The signal from the CH_3 protons is split into a doublet through coupling to the CHO proton, and the CHO signal is a quartet because of coupling to the three CH_3 protons.

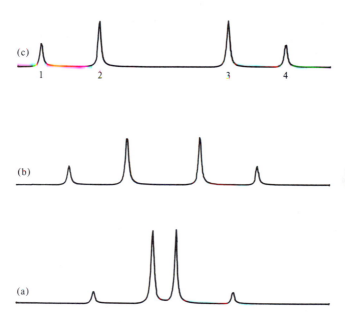

Fig. 6.5 Second-order spectra for two coupled nuclei A and B, both of spin $\frac{1}{2}$. The frequency separation $\nu_A - \nu_B$ is equal to (a) J; (b) $2J$; (c) $3J$. If the frequency separation were negligible in comparison with J, we would obtain a single central line; if it were very large in comparison with J, we would obtain two doublets, with all four lines of equal intensity (see rules 1. and 2. in Section 6.2.2). ν_A and ν_B are the frequencies A and B would generate if there were no spin–spin coupling.

Table 6.1 Frequencies and intensities of the lines in an AB spectrum.

Line number	Frequency	Intensity
1	$\frac{1}{2}J + C$	$1 - \sin 2\theta$
2	$-\frac{1}{2}J + C$	$1 + \sin 2\theta$
3	$\frac{1}{2}J - C$	$1 + \sin 2\theta$
4	$-\frac{1}{2}J - C$	$1 - \sin 2\theta$

C is equal to $\frac{1}{2}\sqrt{[J^2 + (v_A - v_B)^2]}$, and θ is given by $\tan 2\theta = J/(v_A - v_B)$. The pattern is centred about the frequency $(v_A + v_B)/2$.

and relative intensities as the frequency separation between the A and B resonances gradually increases while the spin–spin coupling remains constant. This is the sort of effect that would be expected on gradually increasing the magnetic field, B_0, at which the study is performed.

The frequencies of the lines relative to the centre of the pattern and the intensities of the lines are given in Table 6.1. Note that the splitting between the pair 1 and 2 and between the pair 3 and 4 is always equal to J.

6.2.4 Spin decoupling

The ^{13}C signal from, say, the ^{13}CH$_3$ groups of acetone is split into four components as a result of spin–spin coupling with methyl protons. It can be shown, both theoretically and experimentally, that if irradiation is applied at the resonance frequency of the methyl protons, the ^{13}C quartet is collapsed into a single line. This process of spin decoupling occurs provided that the amplitude of the irradiating field (generally termed B_2) is such that $\gamma B_2 \gg nJ$, where J is the coupling constant and n is the number of lines in the multiplet.

Spin decoupling is very widely used and is of considerable value as a means of both assigning and simplifying spectra. In ^{13}C NMR experiments, for example, it is common practice to use broad-band proton decoupling, i.e. to apply the second radiofrequency field B_2 in such a way that the whole of the ^{1}H spectral range is irradiated. This causes a collapse of many of the ^{13}C multiplets into singlets, which provides extensive simplification of the spectra. However, this is by no means the only reason for using double irradiation, for the B_2 field can also saturate the ^{1}H resonances and thereby lead to large enhancements of the ^{13}C signal intensities through the nuclear Overhauser effect. This effect is discussed in Section 6.4. In the past, ^{31}P studies of tissue metabolism have rarely used double irradiation, but the potential value of double irradiation for such studies is becoming increasingly apparent.

6.2.5 Phase modulation

Under spin-echo conditions, spin–spin coupled signals undergo a process known as phase modulation. This can give rise to inversion of signals, or to much more complex behaviour. As discussed in Section 8.11, this phase modulation can be manipulated by decoupling or by other methods in order to 'edit' spectra, i.e. to detect specific signals of interest.

6.3 RELAXATION

When a sample is placed in a magnetic field it becomes magnetized, because slightly more nuclear spins align with the field than against it (see Chapter 5). At equilibrium the component of magnetization along the field (the z-axis) is equal to M_0, whereas the net magnetization M_{xy} perpendicular to the field is zero. Following any perturbation of this magnetization, e.g. following the application of a 90° pulse, processes take place whereby M_z and M_{xy} return to their equilibrium values of M_0 and zero, respectively. These so-called relaxation processes play a critical role in the theory and practice of NMR.

6.3.1 Spin–lattice relaxation

The return of M_z to its equilibrium value is termed spin–lattice relaxation and is characterized by a time constant T_1 known as the spin–lattice, or longitudinal, relaxation time. The term spin–lattice is used because the processes involve an exchange of energy between the nuclear spins and their molecular framework, which is referred to as the lattice regardless of the physical state of the system.

The return of the nuclear spins to thermal equilibrium, i.e. to equilibrium with the lattice, is often, although not always, exponential. If it is exponential, we can write a simple equation[1] describing the relaxation

$$\frac{\mathrm{d}M_z}{\mathrm{d}t} = \frac{M_0 - M_z}{T_1} \tag{6.4}$$

This equation expresses the return of the magnetization M_z to its equilibrium value M_0 with a time constant T_1.

Let us now consider the nature of the processes that are responsible for relaxation. We have seen that the relative populations of the spin states can be altered in a well-defined way by means of a resonant B_1 field applied in the xy-plane. In a similar manner, *any* fluctuating magnetic field that has a component in the xy-plane that oscillates at the resonant frequency will induce transitions between the spin states of the nuclei. If these fluctuating fields are associated with the lattice, there will be an exchange of energy until the nuclear spins are in thermal equilibrium with the lattice. Thermal

---

[1] This equation and eqn (6.8) are simplified forms of the well-known Bloch equations (see Bloch 1946).

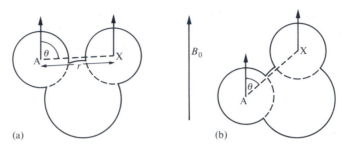

Fig. 6.6 Two nuclei A and X within a molecule both have their spins parallel to B_0. The field at A due to the spin of X is dependent on the angle θ, defined as shown in the diagram. As the molecule rotates from (a) to (b), the angle θ changes, and hence the interaction between A and X also changes. r is the distance between A and X.

equilibrium is characterized by the temperature T of the lattice, and the relative populations of the nuclear energy states at equilibrium (see Section 5.2.4) will be those characteristic of this temperature, i.e.

$$n^-/n^+ = \exp(-\Delta E/kT) \tag{6.5}$$

What processes generate these fluctuating magnetic fields? As an example, consider two nuclei A and X, both of spin $\frac{1}{2}$, within a molecule. Since X has a magnetic dipole moment, it produces a field at A whose component B_{xy} in the xy-plane is $\frac{3}{2}\sin\theta\cos\theta\,\gamma/\hbar r^3$ (see Fig. 6.6, where θ and r are defined). If the molecule is tumbling around in solution, the relative orientations of A and X change randomly. As a result, the angle θ fluctuates, and hence the field B_{xy} also fluctuates. If there are any components of molecular motion that happen to fluctuate at the resonance frequency, then these can cause relaxation of nucleus A. Relaxation generated by this interaction between neighbouring nuclear magnetic dipole moments is termed dipole–dipole relaxation. It should be noted that the two nuclei do not have to be in the same molecule; for example, diffusion of molecules can modulate fields by changing both θ and r, and can therefore also lead to dipole–dipole relaxation.

The frequency distribution of the motion of a randomly tumbling molecule can be expressed in terms of the *spectral density*, $J(\omega)$, and under many circumstances is given by

$$J(\omega) = \frac{\tau_c}{1 + \omega^2\tau_c^2} \tag{6.6}$$

where τ_c is known as the correlation time and expresses the characteristic time-scale of the molecular motion. For example, at room temperature water molecules in solution have a correlation time of about 3×10^{-12} s, ATP might be expected to have a correlation time of about 10^{-10} s, and an enzyme of molecular weight 20000 might have a correlation time of about 10^{-8} s.

We might expect that the relaxation rate $1/T_1$ would be dependent upon the magnitude of the fluctuating fields and upon the spectral density (or components of motion) at the resonance frequency ω_0, and indeed it can be shown that

$$\frac{1}{T_1} \propto B_{xy}^2 \frac{\tau_c}{1 + \omega_0^2 \tau_c^2} \qquad (6.7)$$

The variation of this expression with τ_c is such that, as expected, this relaxation mechanism is most efficient when $\tau_c = 1/\omega_0$, i.e. when the characteristic frequency $(1/\tau_c)$ of the molecular motion is equal to the resonance frequency. However, it should be pointed out that there is an additional contribution to dipole–dipole relaxation which makes the full expression for T_1 rather more complicated (see Section 6.3.3).

6.3.2 Spin–spin relaxation

The return of M_{xy} to its equilibrium value is termed spin–spin relaxation and is characterized by a time constant T_2 known as the spin–spin, or transverse, relaxation time. The term spin–spin is used because the relaxation processes involve interactions between neighbouring nuclear spins without any exchange of energy with the lattice. The equilibrium value of M_{xy} is zero as there is no preferred orientation within the xy-plane, and therefore the equation describing the return to equilibrium may be written

$$\frac{\mathrm{d}M_{xy}}{\mathrm{d}t} = -\frac{M_{xy}}{T_2} \qquad (6.8)$$

This equation expresses the exponential decay of M_{xy} to zero with a time constant T_2.

This decay of M_{xy} is illustrated in Fig. 6.7. Figure 6.7(a) shows the xy-magnetization generated by a 90° pulse. If there were just a single resonance frequency, ω_0, the magnetization would precess coherently about the z-axis with frequency ω_0; in the rotating frame of reference the magnetization would remain static and of constant magnitude. However, if there is a spread of frequencies, $\Delta\omega_0$, different nuclei will precess at slightly different frequencies; in the rotating frame the nuclear magnets will disperse, as shown in Fig. 6.7(b) and (c), the net effect being that M_{xy} decays. The greater the frequency spread, the more rapidly M_{xy} decays. If T_2 represents the time constant of the decay of M_{xy} that results from spin–spin relaxation, then we find that

$$\frac{1}{T_2} = \frac{\Delta\omega_0}{2} = \pi\Delta\nu_{1/2} \qquad (6.9)$$

where $\Delta\nu_{1/2}$ is the corresponding resonance linewidth. Spin–spin relaxation

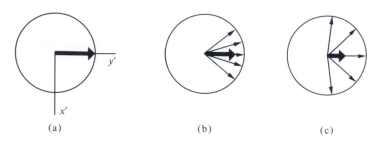

(a)   (b)   (c)

Fig. 6.7 The magnetization in the $x'y'$-plane of the rotating frame. (a) represents the situation immediately after the application of a 90° pulse; the net magnetization is represented by the heavy arrow. The individual nuclear magnets then disperse, as shown by the light arrows in (b), and then (c). Therefore, the *net* magnetization in the $x'y'$-plane gradually declines, the time constant being given by T_2. Additional effects such as B_0 inhomogeneity cause the net magnetization in this plane to decay even more rapidly.

therefore involves processes that cause an inherent broadening of the resonance linewidths.[1] We now consider the nature of these processes.

First, we note that, as a result of spin–lattice relaxation processes, the nuclear spins have a finite lifetime in a given energy state. There is therefore an inherent uncertainty in the resonance frequency, $\Delta\omega_0 \approx 1/T_1$, and therefore an inherent 'lifetime' broadening of the resonances by this amount. Thus, all processes that contribute to spin–lattice relaxation also affect T_2, and it turns out that T_2 cannot be longer than T_1.

In biological NMR, the lifetime broadening is often much less important than the effects we now consider. When there are two neighbouring nuclei A and X, both of spin $\frac{1}{2}$, we have seen that the fluctuating fields in the xy-plane produced by X at A cause spin–lattice relaxation of the A nucleus. Now consider the component of field B_z in the z-direction produced at A by X. The magnitude of B_z is $\frac{1}{2}$ $(3\cos^2\theta - 1)\gamma/\hbar r^3$ (see Fig. 6.6, where θ and r are defined). Thus B_z depends upon the relative orientations of A and X. If A and X are stationary, for example if they are in a powdered solid, then A will resonate at a frequency determined by its fixed orientation in space relative to X. In other molecules within the solid A and X have different relative orientations, and therefore A will resonate at different frequencies. The net result is that the solid produces a very broad resonance, which can be regarded as a summation of a large number of resonances spread over a range of frequencies. The range of frequencies $\Delta\nu$ is proportional to ΔB_z where ΔB_z is the range of B_z fields experienced by A. The value of $\Delta\nu$ might typically be about 5 kHz.

---

[1] We have seen that in practice the linewidths may be further broadened by the effects of field inhomogeneities, which cause the magnetization to decay with a time constant T_2^* (see also Section 6.3.4).

In a solution the motion of the molecules causes the relative orientations of A and X to fluctuate randomly. This motion tends to average the interaction of A and X over all orientations. If there were complete averaging of the interaction, then each nucleus A would experience the same time-averaged field from X and therefore the resonance from A would become a single narrow line. We might expect that averaging of the interaction would become more complete as the speed of the motion increases and, therefore, that the linewidths would decrease as motion increases. This 'motion narrowing' does indeed occur, and it is found that the linewidth in solution is proportional to the correlation time, τ_c, for the motion under consideration. The linewidth is also proportional to the square of the fluctuating field B_z, and so this contribution to spin–spin relaxation can be expressed by the relationship

$$1/T_2 \propto B_z^2 \tau_c. \tag{6.10}$$

A fairly simple derivation of this relationship is given by Slichter (1978).

Note that the expression $J(\omega) = \tau_c/(1 + \omega^2\tau_c^2)$ (see eqn 6.6) simplifies to $J(\omega) = \tau_c$ for $\omega \approx 0$. Thus eqn (6.10) for T_2 is exactly analogous to eqn (6.7) for T_1, except for the following.

1. Equation (6.10) involves the z-component of the fluctuating field rather than the component in the xy-plane.

2. Equation (6.10) involves the spectral density at $\omega \approx 0$, rather than at $\omega = \omega_0$. Thus, the components of motion at low frequencies contribute to T_2 but not to T_1. However, it should be recalled that T_2 also contains a contribution from those components of motion at $\omega = \omega_0$ that also contribute to T_1. This is the lifetime broadening effect that was considered above.

6.3.3 T_1, T_2, and molecular mobility

From the above discussion, we have seen that fluctuating magnetic fields associated with molecular motion are responsible for both spin–spin and spin–lattice relaxation, but that the time-scale of these fluctuations determines their relative contributions to the two types of relaxation. In particular, slow motions contribute only to spin–spin relaxation, whereas components of motion at the resonance frequency contribute to both spin–spin and spin–lattice relaxation. However, the situation is rather more complicated than this, because it is possible for the two nuclei A and X to undergo simultaneous transitions. Both nuclei can undergo transitions in the *same* direction if there is a component of molecular motion at a frequency equal to the *sum* of the Larmor frequencies of the two nuclei. Such a process will contribute to spin–lattice relaxation, and hence, by lifetime broadening, to spin–spin relaxation also. If A and X are like nuclei, they can also undergo spin–spin exchange in which a downward transition of one nucleus is accompanied

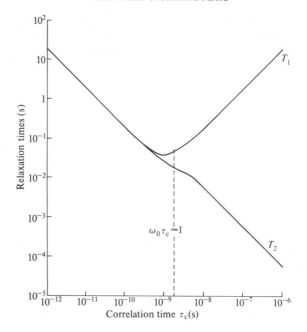

Fig. 6.8 Diagram showing how the relaxation times of the protons in H_2O vary with correlation time τ_c at a resonance frequency of 100 MHz. The graphs are based on eqns (6.11) and (6.12).

by a simultaneous upward transition of the other, without any exchange of energy with the lattice. Such a process does not affect T_1, but it does affect the lifetimes of the nuclear spin states and therefore contributes to T_2. In fact this contribution to T_2 has a similar form to eqn (6.10).

The calculation of T_1 and T_2 is not straightforward, but we can illustrate the above discussion by simply quoting T_1 and T_2 values for the protons of the water molecule. The values are calculated by considering the dipole–dipole interaction between the two protons of each individual molecule and ignoring the effects of protons on other molecule. The results are

$$\frac{1}{T_1} = \frac{3}{10}\frac{\gamma^4\hbar^2}{r^6}\left(\frac{\tau_c}{1+\omega_0^2\tau_c^2} + \frac{4\tau_c}{1+4\omega_0^2\tau_c^2}\right) \qquad (6.11)$$

$$\frac{1}{T_2} = \frac{3}{20}\frac{\gamma^4\hbar^2}{r^6}\left(3\tau_c + \frac{5\tau_c}{1+\omega_0^2\tau_c^2} + \frac{2\tau_c}{1+4\omega_0^2\tau_c^2}\right) \qquad (6.12)$$

The various terms in these expressions correspond to the different types of interaction discussed above. The variation of T_1 and T_2 with correlation time, as deduced from eqns (6.11) and (6.12), is given in Fig. 6.8.

When $\omega_0\tau_c \ll 1$, it can be seen from Fig. 6.8 that $T_1 = T_2$, and the absolute values of T_1 and T_2 are estimated to be about 7 s at room temperature. This is consistent with the measured T_1 and T_2 values for pure water of 3.6 s, as other mechanisms also contribute to the observed relaxation.

When $\omega_0 \tau_c$ is much greater than 1, T_2 becomes much shorter than T_1. A small ratio of $T_2:T_1$ is common in biological NMR; indeed the T_2 values for tissue water tend to be about ten-fold smaller than the T_1 values. Our understanding of water relaxation *in vivo* is far from complete, but it is presumed that there is exchange between free and bound protons, the bound protons having a sufficiently long correlation time to account for the relatively short T_2 values that are observed.

6.3.4 Relaxation mechanisms

In the preceding sections we have considered only one type of relaxation mechanism, namely that generated by the interaction between neighbouring nuclear magnetic dipole moments. In fact, there are many different mechanisms that can contribute to the observed relaxation. If two or more independent relaxation mechanisms exist, the net effect is obtained by adding the individual contributions to the relaxation rates. Thus:

$$1/T_1 = 1/T_{1A} + 1/T_{1B} + 1/T_{1C} + \dots \qquad (6.13)$$

where T_1 is the measured spin–lattice relaxation time, and T_{1A}, T_{1B}, T_{1C}, etc. are the relaxation times that characterize the individual relaxation mechanisms. A similar expression is obtained for the spin–spin relaxation time, i.e.

$$1/T_2 = 1/T_{2A} + 1/T_{2B} + 1/T_{2C} + \dots \qquad (6.14)$$

In view of the inverse relationship with linewidths (see eqn (6.9)), the expression for T_2 is equivalent to the following:

$$\Delta v_{1/2} = \Delta v_{1/2}^A + \Delta v_{1/2}^B + \Delta v_{1/2}^C + \dots \qquad (6.15)$$

where $\Delta v_{1/2}^A$, $\Delta v_{1/2}^B$, $\Delta v_{1/2}^C$, etc. are the characteristic linewidths generated by each relaxation mechanism.

$\Delta v_{1/2}$ is the contribution to signal linewidths that is made by intrinsic relaxation processes. However, we have seen in Section 1.7 that the presence of field inhomogeneities makes an additional contribution, and we can therefore write

$$\Delta v_{obs} = \Delta v_{1/2} + \Delta v' \qquad (6.16)$$

where Δv_{obs} is the observed linewidth and $\Delta v'$ is the contribution to this linewidth that is associated with the field inhomogeneities. Equivalently,

$$1/T_2^* = 1/T_2 + 1/T_2' \qquad (6.17)$$

This equation characterizes the relationship between T_2 and T_2^*, a relationship that is taking on increasing importance in MRI as T_2^*-weighted imaging becomes more widely used for rapid imaging and functional MRI studies.

We shall now discuss the most important relaxation mechanisms and indicate the extent to which they might be expected to contribute to the observed relaxation.

6.3.4.1 The dipole–dipole interaction

This interaction is the most important source of relaxation for many nuclei of spin $\frac{1}{2}$ and can be expected to provide the dominant relaxation mechanism for the majority of ^1H, ^{13}C, and ^{31}P nuclei in biological molecules. Exceptions are most likely to be nuclei that are far removed from any other nuclear spins. This is because the dipole–dipole interaction has a strong $(1/r^6)$ distance dependence. It should be noted that ^1H nuclei have much larger magnetic dipole moments than ^{13}C and ^{31}P nuclei, and therefore the dipole–dipole relaxation tends to be more efficient for protons than for ^{13}C and ^{31}P nuclei.

6.3.4.2 Chemical shift anisotropy

Chemical shift anisotropy, as discussed in Section 6.1.4, describes situations in which the chemical shift of a nucleus depends on the orientation of its molecular environment in the applied field B_0. As the molecule tumbles randomly in solution, the chemical shift and hence the local field will fluctuate, and this local-field fluctuation provides a relaxation mechanism. When there is axial symmetry the contributions of this mechanism to T_1 and T_2 are given by

$$\frac{1}{T_1} = \frac{2}{15} \gamma^2 B_0^2 (\sigma_\parallel - \sigma_\perp)^2 \left(\frac{2\tau_c}{1 + \omega_0^2 \tau_c^2} \right) \tag{6.18}$$

$$\frac{1}{T_2} = \frac{1}{45} \gamma^2 B_0^2 (\sigma_\parallel - \sigma_\perp)^2 \left(4\tau_c + \frac{3\tau_c}{1 + \omega_0^2 \tau_c^2} \right) \tag{6.19}$$

The noteworthy feature of this mechanism is that its effects are proportional to B_0^2. As a result, we can expect the mechanism to become increasingly significant as the field B_0 increases.

Values for $\sigma_\parallel - \sigma_\perp$ often have the same order of magnitude as the chemical shift ranges that characterize the nuclei. For example, $\sigma_\parallel - \sigma_\perp$ is typically a few ppm for ^1H, and 100–300 ppm for ^{19}F, ^{13}C , and ^{31}P. The linewidths of some ^{31}P signals from living systems increase with increasing field strength, and it seems reasonable to interpret at least some of this effect in terms of a significant contribution from chemical shift anisotropy.

An unfortunate consequence of chemical shift anisotropy is that it can significantly reduce the gain in spectral resolution that might be expected on increasing the magnetic field strength. If this mechanism is dominant, the linewidth will increase according to B_0^2, whereas the frequency range of the spectrum increases according to B_0, and therefore the resolution will actually decrease on increasing the magnetic field. The best spectral resolution is obtained at a value of the field such that chemical shift anisotropy accounts for half the linewidth and is not necessarily obtained at the highest field that might be available.

6.3.4.3 *Spin rotation*

When a molecule rotates, the electrons participating in this rotation produce a local magnetic field at the nuclei, the magnitude of which is dependent on the rotational angular momentum of the molecule. Brownian motion causes this angular momentum to fluctuate, and the local field fluctuations thereby generated can cause relaxation. This mechanism of spin rotation is likely to be more significant for small rather than large molecules and is unlikely to produce significant relaxation in large biological molecules.

6.3.4.4 *Scalar coupling*

The spin–spin coupling interaction (see Section 6.2) is alternatively known as the scalar interaction because its strength is independent of the orientation of the molecules within the applied field. Modulation of the magnitude of this interaction can generate relaxation. The fluctuations are usually *slow* on the time-scale we must adopt when considering relaxation, and so the effect contributes far more significantly to spin–spin than to spin–lattice relaxation. Relaxation through scalar coupling is unlikely to be of widespread importance for biological molecules, except in the case of paramagnetic systems where the coupling is between a nucleus and an electron. Such coupling is then referred to as a contact interaction.

6.3.4.5 *Electric quadrupole relaxation*

Nuclei with spin $\frac{1}{2}$ have a spherical charge distribution. This is not true for nuclei of spin greater than $\frac{1}{2}$, and such nuclei possess an electric quadrupole moment. The electric quadrupole moments interact with local electric field gradients. Relaxation is generated by fluctuations in the strength of this interaction resulting from molecular motion. This can be a very efficient relaxation mechanism and it accounts for the broadness of many of the resonances from quadrupolar nuclei.

6.3.4.6 *Relaxation in paramagnetic systems*

Paramagnetic centres generate very powerful relaxation because the magnetic moment of the electron is about 1000 times greater than the nuclear moments. This leads to an enhancement in relaxation by a factor of about 10^6, so even trace amounts of paramagnetic ions can have a profound effect on relaxation times. The effects of paramagnetic species in NMR are discussed in detail in Section 6.5.

6.3.5 Measurement T_1 and T_2

There is an extensive literature, dating back many years, on techniques for the measurement of relaxation times T_1 and T_2. For studies of living systems, there are many reasons why we should wish to make these measurements. For example:

1. In order to select the most suitable radiofrequency pulse interval, it is essential to know the T_1 values of the various signals, or equivalently, their saturation factors (see Section 7.3.3).

2. The measurement of concentrations requires a knowledge of T_1- and T_2-dependent effects on signal intensities.

3. Reaction rate constants can be determined from magnetization transfer studies if the T_1 values of the signals are known (see Section 6.4).

4. Differences in T_1 and T_2 provide the basis for much of the contrast in MRI, and knowledge of T_1 and T_2 values helps in the selection of optimal pulse sequences. Measurements of T_1 and T_2 can also contribute to the characterization and interpretation of imaging findings.

For these and numerous other reasons, measurements of T_1 and T_2 are commonly performed. There are several different radiofrequency pulse sequences that can be used for these measurements, as described in Chapter 8.

6.4 MAGNETIZATION TRANSFER EFFECTS

NMR signals are sensitive to a variety of processes through which magnetization can be transferred from one species to another. In this section, we discuss the two major mechanisms whereby this transfer to take place, namely chemical exchange processes and magnetization transfer through cross-relaxation.

6.4.1 The effects of chemical exchange

Let us consider a simple first-order equilibrium reaction

$$A \underset{k_B}{\overset{k_A}{\rightleftharpoons}} B$$

where k_A and k_B are the first-order rate constants for the reaction. We shall use this reaction to illustrate the ways in which chemical exchange processes can influence the signals of the two species A and B.

Consider first two chemical species A and B of equal concentration which in the absence of exchange would produce two narrow resonances at frequencies ν_A and ν_B. Since the concentrations are equal, we can write $k = k_A = k_B$.

The effects of exchange on the observed spectra are shown in Fig. 6.9. It is convenient to distinguish between three ranges of exchange rates: slow ($k \ll (\nu_A - \nu_B)$), fast ($k \gg (\nu_A - \nu_B)$), and intermediate. When there is slow exchange between the two species (Fig. 6.9(a)), the resonances remain resolved but each is broadened by an amount $\Delta \nu = k/\pi$. This is equivalent to the lifetime broadening we considered in Section 6.3.2; each nucleus only has a lifetime $\tau_m = 1/k$ in a given state and therefore there is an inherent uncertainty equal to k/π in the resonance frequencies.

In fast exchange the nuclei see an average of the A and B environments, and so we obtain a single signal at the average frequency (Fig. 6.9(e)). The width $\Delta \nu$ of this signal decreases as the exchange rate increases (by analogy with the motional narrowing effect on T_2 that was discussed in Section 6.3.2), and is given by

$$\Delta \nu = \pi (\nu_A - \nu_B)^2/2k \qquad (6.20)$$

As the exchange rate increases the averaging process becomes more complete and the line narrows.

In intermediate exchange the spectrum is more complex, as may be seen from Figs 6.9(b)–(e). The separation of the two signals in slow and inter-mediate exchange is given by

$$\delta \nu = [(\nu_A - \nu_B)^2 - 2k^2\pi^2]^{1/2} \qquad (6.21)$$

Therefore the lines gradually merge as the exchange rate increases.

If the concentrations of the two species are not equal, the calculations become much more difficult (see Pople et $al.$ 1959). However, the qualitative features of fast and slow exchange remain similar to those outlined above. For example, when there is fast exchange the observed spectrum consists of a single resonance, with a frequency ν determined by the relative concentrations of the two species:

$$\nu = \frac{[A]\nu_A + [B]\nu_B}{[A] + [B]} \qquad (6.22)$$

Fast exchange is encountered whenever we use the frequency dependence of a resonance to determine pH. Consider, for example, inorganic phosphate, which exists mainly as HPO_4^{2-} and $H_2PO_4^-$ at around neutral pH. The equilibrium between these species can be written

$$H^+ + HPO_4^{2-} \rightleftharpoons H_2PO_4^-$$

In the absence of chemical exchange, the two species would give rise to two resonances separated from each other by about 2.4 ppm. In solution, however, HPO_4^{2-} and $H_2PO_4^-$ exchange with each other rapidly (at about 10^9–10^{10} s^{-1}), and as a result a single resonance is observed, of frequency ν given by eqn (6.22), where A and B represent the two ionization states

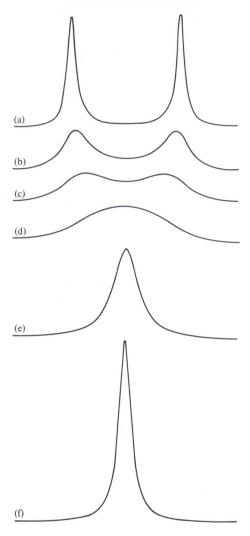

Fig. 6.9 Diagram illustrating the effects of chemical exchange between two species A and B of equal concentration. The exchange rate gradually increases on moving down from spectrum (a) to (f). (a) represents slow exchange, (b)–(e) represent various types of intermediate exchange, and (f) represents fast exchange. The actual rate constants are as follows: (a) $k = \pi(\nu_A - \nu_B)/10$; (b) $k = 3\pi(\nu_A - \nu_B)/10$; (c) $k = \pi(\nu_A - \nu_B)/2$; (d) $k = 4\pi(\nu_A - \nu_B)/5$; (e) $k = 2\pi(\nu_A - \nu_B)$; (f) $k = 4\pi(\nu_A - \nu_B)$.

of inorganic phosphate. Since the relative concentrations of HPO_4^{2-} and $H_2PO_4^-$ are determined by $[H^+]$, the frequency of the observed signal measured as a function of pH gives a 'standard' pH curve (see Section 2.7).

In addition to affecting chemical shifts, linewidths, and lineshapes, exchange also tends to average the spin–lattice relaxation of the two resonances. In general, this will cause the observed spin–lattice relaxation to be non-exponential. However, when species A has a far greater concentration than species B, the measured spin–lattice relaxation of A remains exponential, with a time constant T_{1obs} given by (Luz and Meiboom 1964)

$$1/T_{1obs} = 1/T_{1A} + f/(\tau_B + T_{1B}) \qquad (6.23)$$

where T_{1A} and T_{1B} are the inherent T_1 values of A and B in the absence of exchange, f is the ratio of B to A, and τ_B is the lifetime of the B species. In fast exchange τ_B is much smaller than T_{1B}, and therefore

$$1/T_{1obs} = 1/T_{1A} + f/T_{1B} \qquad (6.24)$$

This equation describes complete averaging of the relaxation rates.

Our discussion so far has assumed that the two signals are very narrow in comparison with their frequency separation. Often, this is not true; in fact, we frequently encounter the situation in which a molecule is exchanging between solution and a bound state. The bound molecule generally produces a much broader signal than the free form, and under these conditions the equation describing the observed lineshape can become very complex. The equation can be simplified if it is assumed that the bound species is present at a much lower concentration than the free form (see Swift and Connick 1962), and it can be simplified even further if there is only a very small frequency difference between the two signals. If these assumptions are valid, the observed spin–spin relaxation time of the free form is given by

$$1/T_{2obs} = 1/T_{2A} + f/(\tau_B + T_{2B}) \qquad (6.25)$$

where A represents the free and B represents the bound form of the molecule. This equation is exactly analogous to the expression for T_1 given above.

This type of analysis is relevant to studies attempting to explain the relaxation characteristics of tissue water in terms of the various compartments of 'free' and 'bound' water that may be present within tissues. Numerous such studies have been undertaken (see Fullerton 1992), but this is a complex area of research because the water protons experience such a diverse range of intracellular environments and interactions.

6.4.1.1 The determination of exchange rates

If exchange rates are very slow, i.e. in the range of minutes to days, they can be measured by applying a chemical perturbation to the system and watching the return to equilibrium. Fast rates, of the order of $100\,s^{-1}$ or more, can have dramatic effects on signals, while intermediate exchange rates in the

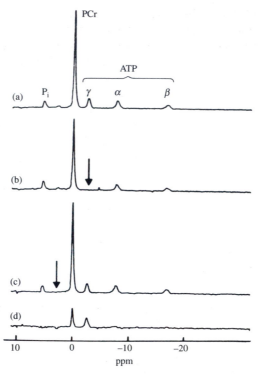

Fig. 6.10 ^{31}P spectra obtained from frog gastrocnemius muscles, illustrating the effects of saturation transfer, as described in the text. The arrows in (b) and (c) indicate the frequencies of the saturating irradiation. (From Gadian *et al*. 1981.)

range 0.1–1 s^{-1} can often be detected by their effects on T_1 and phenomena such as saturation that are related to T_1. In this section, we consider measurements of these intermediate rates, which have proved particularly useful for monitoring the rates of certain enzyme-catalysed reactions *in vivo* (see Sections 2.9 and 3.1). The measurements are achieved using magnetization transfer techniques that involve magnetic perturbation of one species, and observation of how this perturbation influences the magnetization of species with which it is exchanging chemically.

   The spectra of Fig. 6.10 provide an illustration of this type of study. Figure 6.10(a) shows a ^{31}P spectrum obtained from frog gastrocnemius muscles at rest, with the characteristic signals from ATP, phosphocreatine, and inorganic phosphate. Figure 6.10(b) shows a spectrum obtained under similar conditions, except that the signal from the γ-phosphate of ATP was caused to disappear by applying additional irradiation selectively at the frequency of this signal. Effectively, the γ-phosphate is being labelled with zero magnetization. Since the γ-phosphate of ATP is in chemical exchange with the phosphate of phospho-creatine through the reaction catalysed by creatine kinase, some of the label (i.e. loss of signal) is transferred to the phosphocreatine, which therefore has a lower signal intensity in Fig. 6.10(b) than in Fig. 6.10(a). Figure 6.10(c) shows a control spectrum, with irradiation on the opposite side of the PCr peak, while Fig. 6.10(d) represents the difference between the spectra of Figs 6.10(b) and 6.10(c), illustrating the loss of intensity of the PCr signal. From a series of such spectra accumulated under specified conditions, it is possible to measure the rate of interconversion of phosphocreatine and ATP through the creatine kinase reaction. This particular type of magnetization transfer study, in which one of the signals is selectively saturated, is commonly known as saturation transfer.

Here, we briefly outline the theory of this approach, which was discussed in depth by Forsen and Hoffman (1963) some years ago.

Consider the equilibrium reaction

$$A \underset{k_B}{\overset{k_A}{\rightleftharpoons}} B$$

where k_A and k_B are the first-order rate constants for the forward and back reactions, respectively. Suppose that the equilibrium values of the magnetization of A and B (which in the above example correspond to phosphocreatine and ATP respectively) are M_{0A} and M_{0B} respectively, and that the z-components of magnetization at any given time are M_A and M_B. In the absence of exchange between A and B, M_A would return to its equilibrium value M_{0A} with a time constant T_{1A}, and we could write

$$\frac{dM_A}{dt} = \frac{M_{0A} - M_A}{T_{1A}} \qquad (6.26)$$

In the presence of exchange, this equation must be modified as A loses magnetization through conversion to B and gains magnetization from the reverse process. The full equation under these conditions is

$$\frac{dM_A}{dt} = \frac{M_{0A} - M_A}{T_{1A}} - k_A M_A + k_B M_B \qquad (6.27)$$

Suppose now that the resonance of B is selectively saturated so that $M_B = 0$. Equation (6.27) then becomes

$$\frac{dM_A}{dt} = \frac{M_{0A} - M_A}{T_{1A}} - k_A M_A \qquad (6.28)$$

In the steady state $dM_A/dt = 0$, and therefore

$$0 = \frac{M_{0A} - M_A}{T_{1A}} - k_A M_A$$

from which

$$M_A = \frac{M_0}{1 + k_A T_{1A}} \qquad (6.29)$$

i.e. the magnetization of A in the steady state is no longer equal to M_{0A}, but is equal to M_{0A} reduced by a factor $(1 + k_A T_{1A})$.

The measurement of k_A using eqn (6.29) requires knowledge of the ratio M_A/M_0 and of T_{1A}, and can be carried out as follows. Consider experiments in which radiofrequency pulses are applied every TR seconds. If TR is much greater than T_{1A}, the magnetization of A reaches a steady state prior to application of each pulse, and so a comparison of the observed signal intensity in the presence and absence of selective irradiation of B yields the magnitude of $(1 + k_A T_{1A})$. To determine k_A, it is then necessary to measure T_{1A}. Unfortunately, T_{1A} cannot be measured from a conventional T_1 experiment because T_{1A} represents the inherent spin–lattice relaxation time of A that would be observed in the absence of chemical exchange. Any measurement made in the presence of exchange must necessarily include a contribution from the relaxation of B. This contribution can be evaluated using the theory developed by Forsen and Hoffman (1963), but there is an alternative, simpler method of obtaining the required information.

The full solution to eqn (6.27) shows that, when B is saturated, M_A returns to its steady-state value $M_{0A}/(1 + k_A T_{1A})$ with a time constant $T_{1\mathrm{eff}}$ given by $1/T_{1\mathrm{eff}} = 1/T_{1A} + k_A$. Therefore, a T_1 measurement made in the presence of selective irradiation of B enables k_A and T_{1A} to be determined from the simultaneous equations

$$M_A = M_{0A}/(1 + k_A T_{1A}) \qquad (6.30)$$

$$\frac{1}{T_{1\mathrm{eff}}} = \frac{1}{T_{1A}} + k_A \qquad (6.31)$$

Similar measurements can be made for the determination of k_B and T_{1B}. It should be pointed out that several types of experimental procedure are available for determining the rate constants. For example, the B resonance may be perturbed by a selective inversion pulse, rather than by saturation; this produces more extensive transfer of magnetization, although the analysis becomes somewhat more complex. Further information about the technique of magnetization transfer and its applications to tissue bioenergetics is given in a series of articles on cardiac muscle (see Ugurbil 1985; Ugurbil et al. 1986; Kingsley-Hickman et al. 1987) and in a review by Brindle (1988).

6.4.2 Magnetization transfer by cross-relaxation

In the above section, we described how chemical exchange processes can mediate the exchange of magnetization between different nuclear species. Magnetization may also be transferred in the absence of any chemical exchange, as a result of processes involving cross-relaxation. In this section we briefly discuss the nature of these processes, and the ways in which they can be exploited.

6.4.2.1 *Applications in MRS*

We have seen in Section 6.3.4 that the relaxation for many nuclei is dominated by dipole–dipole interactions between neighbouring nuclear spins. For example, protonated ^{13}C nuclei are relaxed largely through this type of interaction with neighbouring protons. Under normal conditions the relative populations of the protons in their two spin states are given by the Boltzmann distribution (see Section 5.2.4). It can be shown that if this population distribution is altered, for example by saturating the proton spins, then the dipole–dipole interaction with the neighbouring ^{13}C nuclei causes a change to take place in the relative populations of the ^{13}C spin states. This results in a change in the steady-state magnetization of the ^{13}C nuclei, and is known as the nuclear Overhauser effect. It can be shown that saturation of the ^{1}H resonances (which corresponds to the populations of the spin states being equalized) can lead to an enhancement in the intensity of the *observed* signals (in this case the ^{13}C signals) given by the expression

$$\frac{\text{signal intensity}\,(^{1}\text{H saturated})}{\text{signal intensity}\,(\text{no saturation})} = 1 + \frac{\gamma_s}{2\gamma_0} \qquad (6.32)$$

where γ_s and γ_0 are the gyromagnetic ratios of the saturated and observed nuclei respectively. However, this equation is only valid if the spin–lattice relaxation of the observed nuclei results entirely from dipole–dipole coupling, and also if the motion is sufficiently rapid for the condition $\omega_0\tau_c \ll 1$ to be satisfied. If either of these criteria is not fulfilled, the enhancement is decreased; indeed, when the motion is slow, the observed signal intensity may actually decrease rather than increase.

In ^{13}C NMR experiments the nuclear Overhauser effect often produces a significant intensity enhancement, and therefore proton irradiation is commonly applied in order to obtain this important improvement in the signal-to-noise ratio. For studies of molecular structure, the degree of enhancement can be a powerful aid to conformational analysis because the magnitude of the dipolar coupling is strongly distance dependent (proportional to $1/r^6$). It should be noted that proton irradiation can also simplify spectra and further improve the signal-to-noise ratio by means of spin decoupling (see Sections 6.2.4 and 8.11).

The nuclear Overhauser effect is particularly useful in ^{15}N NMR. This is because the ^{15}N nucleus has a small gyromagnetic ratio, and from eqn (6.32) it can be seen that this leads to a large enhancement. It is interesting to note that the gyromagnetic ratio of the ^{15}N nucleus is negative, and therefore in general the enhancement factor is also negative. It is for this reason that many of the signals in ^{15}N NMR spectra appear inverted.

The nuclear Overhauser enhancement for ^{31}P nuclei is usually smaller than for ^{13}C nuclei, partly because the gyromagnetic ratio for ^{31}P nuclei is larger and partly because ^{31}P nuclei are rarely bonded directly to protons. In addition, the ^{31}P–^{1}H spin–spin coupling constants are relatively small, and so there is not as much to be gained by spin decoupling. For these reasons, it has not been so common to employ proton irradiation in biological ^{31}P NMR experiments. However, recent studies (Bottomley and Hardy 1992; Murphy-Boesch et al. 1993) have shown that there are indeed significant benefits in using proton irradiation for ^{31}P studies, particularly at relatively low field strengths (1.5 T) where the unresolved ^{31}P–^{1}H spin–spin couplings add significantly to the broadness of the phosphomonoester and phosphodiester signals. As a result, it appears likely that proton irradiation will become increasingly used for ^{31}P MRS studies, as well as for ^{13}C MRS. The use of proton irradiation, both for the nuclear Overhauser enhancement and for spin decoupling, is discussed further in Section 8.11.

6.4.2.2 Applications in MRI

Cross-relaxation effects, of a similar nature to those discussed in the above section, are also believed to be responsible for the phenomenon of magnetization transfer contrast in MRI, in which a decrease in the water signal is observed following off-resonance irradiation (Wolff and Balaban 1989; Balaban and Ceckler 1992). This phenomenon can be explained in terms of interactions between two spin systems, namely free water protons and macromolecular protons. While the macromolecular protons generate signals that are too broad to detect directly in conventional MRI studies, these broad signals can be selectively saturated using irradiation that is several kHz removed from the water resonance (Fig. 6.11). Saturation of the macromolecular protons results in a perturbation in the steady-state magnetization of the free water protons, as a result of magnetization transfer processes between the two spin systems. In practice, a reduction in water signal is observed, which contrasts with the enhancement of signal intensity that is commonly observed through the nuclear Overhauser effect. The explanation for this comes from a consideration of the equations describing the transfer of magnetization that occurs through dipole–dipole coupling. It was mentioned in Section 6.4.2.1 that the signal enhancement given by eqn (6.32) only holds true if the condition $\omega_0 \tau_c \gg 1$ is satisfied, and that when the motion is slow, the observed signal intensity may actually decrease rather than increase. The slow motion condition is evidently responsible for the reduction in water signal intensity that is seen in these magnetization transfer studies. In fact, the presence of surface hydroxyl and/or amine groups on the macromolecules appears to be necessary for magnetization transfer. It is presumed that these groups orient water protons in close proximity to macromolecular protons for a time sufficient to allow magnetization transfer to occur by dipole interactions. This residence time has been estimated to be of the order of microseconds, giving rise to the condition $\omega_0 \tau_c \gg 1$ and hence to signal loss

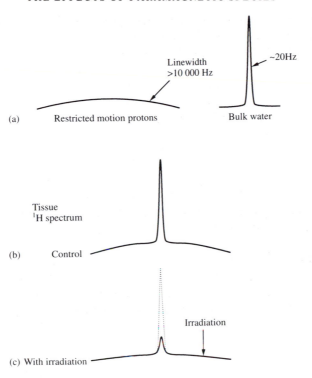

Fig. 6.11 Schematic diagram showing magnetization transfer between proteins with restricted motion (H_r) and bulk water (H_f). (a) shows the individual lineshapes of the two species; (b) shows the two signals together, as they might be observed in tissue; and (c) shows the reduction in the signal from H_f on saturation of the signal from H_r. This reduction is due to magnetization transfer between the two populations. (Reproduced with permission of Raven Press Ltd, New York, from Balaban and Ceckler (1992), *Margn. Reson. Quart.*, **2**, 116–37.)

rather than enhancement (see Scholz *et al.* 1993). Magnetization transfer studies of model protein solutions (Koenig *et al.* 1993) appear to be completely consistent with these findings.

6.5 THE EFFECTS OF PARAMAGNETIC SPECIES

Paramagnetic species such as transition-metal and lanthanide ions have unpaired electron spins, and as a result they have magnetic moments of the order of 1000 times stronger than nuclear magnetic moments. An important consequence of this is that they can strongly influence the characteristics of NMR signals. In this section we briefly describe the main effects of paramagnetics on NMR signals, and outline the ways in which these effects can be exploited in both imaging and spectroscopy.

Paramagnetic centres can enhance relaxation rates and also generate frequency shifts. There is an extensive literature, dating back many years, on the theory underlying the shifts and relaxation effects generated by paramagnetic centres, and the effects of paramagnetic ions in particular are well understood. The use of paramagnetic ions as agents for enhancing the relaxation of water protons was appreciated from the early days of NMR (Bloch 1946), and shortly afterwards their effects were considered theoretically as an extension of the relaxation theory for pure water (Bloembergen *et al.* 1948). The theory was developed further in the 1950s and the Solomon–Bloembergen equations (Solomon 1955; Bloembergen 1957) emerged as the accepted description of the relaxation generated by paramagnetic ions. Paramagnetic species became widely used as relaxation and shift reagents in solution, and when studies of living systems became feasible, it soon became apparent that they could also play a useful role *in vivo*.

It turns out that some paramagnetic ions are particularly efficient enhancers of relaxation, whereas others produce significant shifts with relatively little relaxation effect. Thus there are 'shift reagents' such as Co^{2+}, Ni^{2+}, and most of the lanthanide ions (including Dy^{3+}), and 'relaxation agents' such as Mn^{2+} and Gd^{3+}. The extent to which a given metal ion shifts or relaxes NMR signals depends on a number of factors, including its electron–spin relaxation time, which tends to be long for relaxation agents and short for shift reagents. We begin by discussing shift effects.

6.5.1 Shift effects

Paramagnetic centres can influence resonance frequencies through specific magnetic interactions with neighbouring nuclei, or through relatively non-specific magnetic susceptibility effects.

The specific interactions with neighbouring nuclei can be described in terms of contact or pseudo-contact shifts. Contact shifts involve the transfer of unpaired electron density to a nucleus, an effect that is usually transmitted through chemical bonds. Pseudo-contact or dipolar shifts result from a dipolar interaction between the electron spin and the nuclear spin. Remarkable shifts can be generated; for example, it was mentioned in Section 2.12 that the paramagnetic Fe^{2+} centre of deoxymyoglobin shifts the proximal histidyl NH signal to about 75 ppm.

The non-specific interactions result from magnetic susceptibility effects. Any substance acquires magnetization when placed in a magnetic field, as a result of interactions between its electrons and the field. The magnetic susceptibility χ is a measure of this magnetization, and it differs in sign according to whether the substance is diamagnetic or paramagnetic. In a diamagnetic substance, the induced magnetization opposes the applied field, and χ is negative; in contrast, for paramagnetic substances χ is positive. All substances show a diamagnetic effect (see Section 6.1.2), whereas paramagnetism is associated only with those molecules or ions that possess permanent electronic magnetic moments. If there is a paramagnetic contribution to χ it can far outweigh the diamagnetic contribution. As stated in Section 6.1.2, the effects of this magnetization are to modify the field experienced by a nucleus by an amount B_{bs}

$$B_{bs} = S_f \chi B_0$$

where S_f is a numerical factor that depends on the orientation and shape of the sample.

The frequency shifts generated by paramagnetic species can be exploited in several ways. For example, it was mentioned in Section 2.8.1 that shift reagents have been designed for shifting the extracellular ^{23}Na signal away from the intracellular signal. Dysprosium (Dy^{3+}) complexes are commonly used as shift reagents, partly because Dy^{3+} has a large magnetic moment which generates correspondingly large shifts, and also because its relaxation effects through dipole–dipole interactions are relatively modest. If we consider, for example, the reagent Dy-TTHA, this will generate a specific dipolar shift in the extracellular ^{23}Na signal, as a result of binding of Na^+ ions to the DY-TTHA complex, which is confined to the extracellular compartment. Superimposed upon this, there will be more general shifts due to the magnetic susceptibility effects generated by this paramagnetic agent. The problem with the susceptibility effects is that they are dependent on the shape and orientation of the individual compartments within the sample, and as such they are somewhat unpredictable. In particular, they may cause broadening of spectral lines due to inhomogeneities that are difficult to shim out (see Fig. 2.13), and they may even produce multiple peaks that might be erroneously attributed to different compartments. These undesirable susceptibility effects (in addition to potential toxic effects) have inhibited the more widespread use of paramagnetic agents as shift reagents *in vivo*. However, as discussed below, such effects can be put to advantage in imaging studies, for they generate changes in T_2 or T_2^* that can be used as an important contrast mechanism.

6.5.2 Relaxation effects

The use of paramagnetic ions as contrast agents in MRI is based largely on the relaxation enhancement that they generate, and in this section we discuss the basis of this enhancement.

As discussed in Section 6.3, the dominant relaxation mechanism for the protons of pure water involves dipole–dipole interactions between neighbouring water protons. Since paramagnetic ions have magnetic moments that are about 1000 times greater than the nuclear moments, they produce correspondingly large local fields and can therefore enhance the relaxation rates of water protons in the vicinity of the ions. The closest protons are those of the water molecules that bind directly to the paramagnetic ions. Of course, only a small fraction of the water will be bound to the ions. However, since this fraction continually exchanges with the free water, the effects of the paramagnetic ions are distributed throughout the bulk of the water. There can also be direct relaxation effects experienced by water molecules that are close to, but not bound to, the paramagnetic species; these effects are sometimes termed 'outer-sphere' relaxation. There have been numerous reviews devoted to the relaxation that is generated by paramagnetic ions (see, for example, Watson *et al.* 1992 and refs therein); here we briefly discuss the main factors that influence relaxation enhancement.

The increase in proton relaxation rate generated by paramagnetic ions is directly proportional to the concentration of the ions and to the square of their magnetic moment, but inversely proportional to r^6, where r is the distance between the paramagnetic ion and the bound protons. Thus a small increase in the interatomic distance can cause a dramatic reduction in relaxation effects. Enhancement of relaxation is also dependent on the correlation time characterizing the magnetic interactions and on the resonance frequency. The correlation time, in turn, depends on molecular motion, on the relaxation rate of the electron spins, and on the exchange rate between free and bound water.

On the basis of these relationships, the following considerations enter into the selection of useful contrast agents.

1. To produce strong relaxation effects, the paramagnetic ion should have a high spin number and a long electron–spin relaxation time. Mn^{2+} and Gd^{3+} have fairly high spin numbers ($\frac{5}{2}$ and $\frac{7}{2}$, respectively), and their long electron–spin relaxation times results in these ions being particularly efficient enhancers of relaxation.

2. The distance between the paramagnetic ion and the protons it relaxes should be minimized, i.e. the paramagnetic ion, whether or not it is com-plexed, should bind directly to the water, and preferably each ion should bind to several water molecules. The $1/r^6$ dependence makes this an important con-sideration in the design of paramagnetic complexes. However, if water *is* totally excluded from direct access to the paramagnetic ion, there can nevertheless be relaxation through outer-sphere effects, and these can also be useful. Any bound water must be able to exchange rapidly with the free water in order to distribute the effects of the paramagnetic agent throughout the water.

3. As a result of the dependence of relaxation on correlation time, it is generally preferable to retard molecular motion. For this reason, large complexes should be used, if possible.

4. Complexes of ions with chelating agents such as DTPA are safer than free ions alone. The use of complexes can also overcome problems associated with the poor solubility of some free metal ions at physiological pH.

5. There is considerable interest in the development of more sophisticated contrast agents that could be directed to specific sites. However, to produce useful enhancement of relaxation, an effective paramagnetic concentration of the order of 100 μM or more is required in the region of interest. (This value is very approximate because many factors influence the efficacy of a contrast agent.) This relatively high concentration presents problems in the development and implementation, for example, of antibodies labelled with paramagnetic agents.

6.5.2.1 Relaxation times vs relaxation rates

Although the relaxation characteristics of images are generally expressed in terms of the relaxation *times* T_1 and T_2, the effects of contrast agents are calculated in terms of the relaxation *rates* $1/T_1$ and $1/T_2$. This reflects the fact that, if two or more independent relaxation mechanisms exist, the overall effect is obtained from the addition of the individual relaxation *rates* (see Section 6.3.4). Thus the observed relaxation times $T_{1\text{obs}}$ and $T_{2\text{obs}}$ in the presence of a contrast agent are given by

$$1/T_{1\text{obs}} = 1/T_{1\text{o}} + 1/T_{1\text{p}} \qquad (6.33)$$

$$1/T_{2\text{obs}} = 1/T_{2\text{o}} + 1/T_{2\text{p}} \qquad (6.34)$$

where $T_{1\text{o}}$ and $T_{2\text{o}}$ are the intrinsic T_1 and T_2 values measured in the absence of the agent and $1/T_{1\text{p}}$ and $1/T_{2\text{p}}$ are the contributions that the contrast agent makes to the relaxation rates. Using these equations, it may readily be shown that the observed effects of a contrast agent depend on the intrinsic relaxation measured in the absence of the agent. For example, if $1/T_{1\text{p}}$ were equal to $1\,\text{s}^{-1}$, then an intrinsic T_1 value of 1 s would be reduced by the agent to 500 ms, whereas an intrinsic T_1 value of 100 ms would be reduced only to 91 ms, a much smaller absolute and proportional effect. Indeed, this explains why the predominant effect of contrast agents such as Gd-DTPA tends to be on T_1 rather than T_2; intrinsic tissue T_2 values tend to be much lower than T_1 values, and the contrast agent therefore exerts a smaller absolute and proportional effect on T_2 than on T_1.

6.5.2.2 Additional relaxation and line-broadening effects

Paramagnetic species tend to have adverse effects in metabolic studies; even if they are present in just trace quantities, they may broaden spectral lines significantly. Particularly strong effects may be seen in ^{31}P studies since paramagnetic ions bind tightly to many phosphates, including ATP. These effects are often noticeable in high-resolution studies of tissue extracts, and it is commonly desirable, especially for ^{31}P spectroscopy of tissue extracts, to add an agent such as EDTA to chelate and effectively remove any paramagnetic ions.

Problems with line broadening in spectroscopy can be generated not just by these specific relaxation effects, but also by the effects of magnetic susceptibility variations, as discussed in Section 6.5.1. Susceptibility effects can, however, be exploited in imaging. For example, the presence of paramagnetic species within blood vessels results in local field inhomogeneities in and around the vessels, which in turn can cause signal losses in T_2^*-weighted images. The image contrast generated in this way can be utilized for perfusion and functional imaging studies, as discussed in detail in Chapter 4. Other potential applications include the investigation of abnormalities in brain iron (Ordidge *et al*. 1994).

REFERENCES

Balaban, R. S. and Ceckler, T. L. (1992). Magnetization transfer contrast in magnetic resonance imaging. *Magn. Reson. Quart.*, **2**, 116–37.

Bloch, F. (1946). Nuclear induction. *Phys. Rev.*, **70**, 461–74.

Bloembergen, N. (1957). Proton relaxation times in paramagnetic solutions. *J. Chem. Phys.*, **27**, 572–3.

Bloembergen, N., Purcell, E. M., and Pound, R. V. (1948). Relaxation effects in nuclear magnetic resonance absorption. *Phys. Rev.*, **73**, 679–712.

Bottomley, P. A. and Hardy, C. J. (1992). Proton Overhauser enhancements in human cardiac phosphorus NMR spectroscopy at 1.5 T. *Magn. Reson. Med.*, **24**, 384–90.

Brindle, K. M. (1988). NMR methods for measuring enzyme kinetics *in vivo*. *Prog. NMR Spectrosc.*, **20**, 257–93.

Forsen, S. and Hoffman, R. A. (1963). Study of moderately rapid chemical exchange reactions by means of nuclear magnetic double resonance. *J. Chem. Phys.*, **39**, 2892–901.

Fullerton, G. (1992). Physiological basis of magnetic relaxation. In *Magnetic resonance imaging* (eds D. D. Stark and W. G. Bradley, jun.), pp. 88–108. Mosby Year Book, St. Louis.

Gadian, D. G., Radda, G. K., Brown, T. R., Chance, E. M., Dawson, M. J, and Wilkie, D. R. (1981). The activity of creatine kinase in frog skeletal muscle studied by saturation-transfer nuclear magnetic resonance. *Biochem. J.*, **194**, 215–28.

Kingsley-Hickman, P. B., Sako, E. Y., Mohanakrishnan, P., Robitaille, P. M., From, A. H., Foker, J. E., *et al*. (1987). ^{31}P NMR studies of ATP synthesis and hydrolysis kinetics in the intact heart myocardium. *Biochemistry*, **26**, 7501–10.

Koenig, S. H., Brown, R. D., III, and Ugolini, R. (1993). A unified view of relaxation in protein solutions and tissue, including hydration and magnetization transfer. *Magn. Reson. Med.*, **29**, 77–83.

Luz, Z. and Meiboom, S. (1964). Proton relaxation in dilute solutions of Co(II) and Ni(II) ions in methanol and the rate of methanol exchange in the solution sphere. *J. Chem. Phys.*, **40**, 2686–92.

Murphy-Boesch, J., Stoyanova, R., Srinivasan, R., Willard, T., Vigneron, D., Nelson, S., *et al*. (1993). Proton-decoupled ^{31}P chemical shift imaging of the human brain in normal volunteers. *NMR Biomed.*, **6**, 173–80.

Ordidge, R. J., Gorell, J. M., Deniau, J. C., Knight, R. A., and Helpern, J. A. (1994). Assessment of relative brain iron concentrations using T_2-weighted and

T_2^*-weighted MRI at 3 Tesla. *Magn. Reson. Med.*, **32**, 335–41.

Pople, J. A., Schneider, W. G., and Bernstein, H. J. (1959). *High resolution nuclear magnetic resonance*. McGraw-Hill, New York.

Roberts, G. C. K. (ed.) (1993). *NMR of macromolecules. A practical approach*. Oxford University Press, Oxford.

Scholz, T. D., Ceckler, T. L., and Balaban, R. S. (1993). Magnetization transfer characterization of hypertensive cardiomyopathy: significance of tissue water content. *Magn. Reson. Med.*, **29**, 352–7.

Slichter, C. R. (1978). *Principles of magnetic resonance*, 2nd edn. Springer-Verlag, Berlin.

Solomon, I. (1955). Relaxation processes in a system of two spins. *Phys. Rev.*, **99**, 559–65.

Swift, T. J. and Connick, R. E. (1962). N.M.R. relaxation mechanisms of ^{17}O in aqueous solutions of paramagnetic cations and the lifetime of water molecules in the first coordination sphere. *J. Chem. Phys.*, **37**, 307–20.

Ugurbil, K. (1985). Magnetization-transfer measurements of individual rate constants in the presence of multiple reactions. *J. Magn. Reson.*, **64**, 207–19.

Ugurbil, K., Petein, M., Maidan, R., Michurski, S., and From, A. H. L. (1986). Measurement of an individual rate constant in the presence of multiple exchanges: application to myocardial creatine kinase reaction. *Biochemistry*, **25**, 100–7.

Watson, A. D., Rocklage, S. M., and Carvlin, M. J. (1992). Contrast agents. In *Magnetic resonance imaging* (eds D. D. Stark and W. G. Bradley, jun.), pp. 372–437. Mosby Year Book, St. Louis.

Wolff, S. D. and Balaban, R. S. (1989). Magnetization transfer contrast (MTC) and tissue water proton relaxation in vivo. *Magn. Reson. Med.*, **10**, 135–44.

7

Instrument design and operation

The main units that make up an NMR instrument are:

(1) the magnet and shim coils,

(2) the gradient system,

(3) the transmitter,

(4) the radiofrequency coil(s),

(5) the receiver, and

(6) the computer.

In this chapter we discuss each of these units and the ways in which they interact with each other. In addition, an explanation is given of some of the common operational procedures that are involved in obtaining images or spectra. For simplicity, many of the procedures are introduced in terms of the acquisition of spectra. However, we have seen that imaging and spectroscopy have many features in common; both NMR approaches involve the collection and display of signals as a function of frequency, the main technical difference being that in spectroscopy the frequency encodes for chemistry, whereas in imaging the use of linear field gradients causes the frequency to encode for space. Therefore most of the procedures that are described are equally applicable to the acquisition of images.

7.1 THE MAGNET AND SHIM COILS

7.1.1 The types of magnet

The capability of an NMR instrument is critically dependent upon the magnitude and homogeneity of the static magnetic field and on the bore size of the magnet. There are three main types of magnet; permanent, resistive, and superconducting. Permanent magnets make use of ferromagnetic materials. They do not require an external current supply, but for various reasons, including the saturation field strength and the tonnage of material required, their main use is in relatively low-cost clinical imaging systems operating at low field strengths. They have the advantage of providing an open magnet system with easy patient access.

Resistive magnets are electromagnets in which the field is generated by means of currents through conventional electric conductors. Owing to the resistance of such conductors, there are substantial power requirements, which increase with field strength. As a result, these magnets are also restricted to relatively low field strengths. Open resistive systems with easy patient access are now available.

Permanent and resistive magnets are used in clinical imaging systems that operate at field strengths in the range 0.05–0.4 T. Higher field clinical systems require superconducting magnets. Superconductors are materials such as niobium–titanium alloys that lose all their resistance at very low temperatures. To achieve a sufficiently low temperature, the conductors need to be enclosed in a Dewar vessel of liquid helium. The boil-off of liquid helium is minimized by means of surrounding vacuums, radiation shields, and (in many systems) liquid nitrogen chambers, and this complete cooling system forms a major component of the magnet design. Superconducting magnets are initially energized by means of an external power supply. Once the current has reached the required level, a resistive switch across the ends of the magnet is allowed to become superconducting, the power supply is disconnected, and then currents continue to circulate through the superconductor with no dissipation of energy and minimal drift. Many kilometres of superconducting windings are required in order to generate magnetic fields of the requisite strength and homogeneity. In the next section, we discuss these homogeneity and field-strength requirements, but before doing so it is worth emphasizing that the continued push towards more homogeneous, higher field superconducting magnets was initiated primarily by chemists and biochemists wishing to use NMR spectroscopy to investigate molecules in solution. For such studies, vertical magnets with field strengths of up to 17.5 T are now available (for comparison, the Earth's magnetic field is about 5×10^{-5} T). However, the narrow bore size of these magnets imposes limits on the range of metabolic and imaging studies that can be carried out. In practice, therefore, a wide range of magnet systems are available. These include horizontal wide-bore magnets for clinical investigations, higher field horizontal magnets with narrower bores for animal studies, and very high field vertical magnets for studies of small samples.

7.1.2 Field homogeneity

We saw in Section 1.7 that a critical feature of NMR magnets, particularly for spectroscopy studies, is that the field strength should be extremely homogeneous over the sample volume. For high-field studies of molecules in solution, field homogeneity as remarkable as 1 part in 10^9 is required in order to achieve linewidths of less than 1 Hz at frequencies of 300–600 MHz. For clinical studies, the criteria are not so stringent, but nevertheless ^{1}H MRS studies of the brain require homogeneity of better than 5 parts in 10^8 if linewidths of 3 Hz at 63 MHz (i.e. 1.5 T) are to be obtained. For imaging too, good homogeneity is important. Although the requirements are not so striking as for spectroscopy studies, the increasing use of gradient echo imaging is

7.1.3 The choice of field strength

It is generally considered desirable to perform NMR studies at the highest magnetic field that is available, for an increase in field strength will normally increase both signal-to-noise ratios and spectral resolution. A high field is certainly desirable for solution studies, and indeed for almost all experiments utilizing relatively small amounts of material (a few ml or less). Therefore studies of body fluids, tissue extracts, and cellular suspensions are best performed on high-field systems in the range of 7 T and above. However, for *in vivo* studies, there are additional criteria that must be considered.

First, there are practical difficulties with building very high field magnets that are large enough to accommodate human subjects. As a result, magnets that are available for clinical studies are presently limited to field strengths of 4 T or below, and the costs and siting problems of 4 T systems are such that only very few of these magnets are in use. Very high field systems also generate additional problems with respect to radiofrequency power requirements, radiofrequency penetration into the body, and the more general question of safety (see Kanal *et al.* 1990; Shellock and Kanal 1994 and refs therein for safety considerations in MRI). The recent developments in functional neuroimaging, together with continued progress in MRS, have created considerable interest in a new generation of 3 T systems, and there is little doubt that, at least for research purposes, 3 T clinical systems will proliferate in the coming years. Most clinical studies, however, are carried out at field strengths of 1.5 T and below. A field strength of 1.5 T is generally regarded as the minimum at which spectroscopy can be reasonably performed, although a number of 1.0 T systems are being used with some success for ^{1}H MRS of the brain. MRI can be, and indeed is, successfully carried out over a wide range of field strengths; commercial systems are available to field strengths as low as 0.05 T.

While it is tempting to comment on *optimal* field strengths for studies of living systems, it is not straightforward to come to any definitive conclusions, as there are so many factors that need to be considered. For example, spectra obtained at very high field strengths may, in some cases, suffer from the effects of chemical shift anisotropy (see Section 6.3), which can significantly reduce the gain in spectral resolution that might be anticipated on increasing the field strength. Nevertheless, the comments above provide general guidelines for the most appropriate field strengths for various types of study.

7.2 THE GRADIENT SYSTEM

The generation of magnetic resonance images (and of most localized spectra) relies on the appropriate use of pulsed magnetic field gradients, and in this section we discuss the requirements for these gradients. These are best

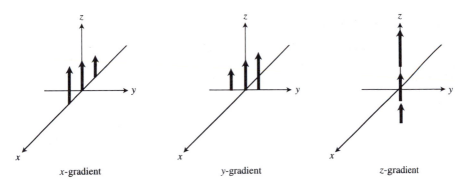

x-gradient   y-gradient   z-gradient

Fig. 7.2 Diagram illustrating the field gradients generated in each of the x-, y-, and z-directions.

considered by reference to a typical two-dimensional Fourier-transform imaging sequence (see Fig. 5.14), which requires gradients in each of the x-, y-, and z-directions. The x-gradient, G_x, generates a field directed along the z-axis (i.e. parallel to B_0) which varies linearly with the x-coordinate, as shown in Fig. 7.2 (it does *not* generate a field oriented along the x-direction). Thus the resonance frequency in the presence of the G_x gradient varies linearly with x. Similarly, the G_y gradient produces a field directed along the z-axis which varies linearly with the y-coordinate. These gradients are generated in just the same way as those produced by the shim coils, i.e. by specially constructed coils mounted within the bore of the magnet, designed to produce field gradients of the required strength and linearity.

What are the requirements for such coils? The first requirement is to produce gradients with a high degree of linearity, for it may readily be shown that non-linearity produces distortion of images. In practice, non-linearity is not normally a problem, except towards the edges of images with very large fields of view, at relatively large distances from the centre of the magnet. Secondly, a lot of gradient switching is required for each data acquisition. Much of this sequence of events needs to take place within a time that is shorter than or comparable to T_2 (typically about 100 ms or less), and so there is a requirement for rapid gradient switching, in order that full power can be reached in less than 1 ms. Thirdly, the gradients need to be sufficiently strong to generate the required spatial resolution. Let us briefly consider the typical needs.

For slice selection, the gradient is applied in the presence of a frequency-selective radiofrequency pulse (see Section 5.7.2). Typically, a pulse will have a duration in the range 1–10 ms, with a bandwidth of, say, 1 kHz. If a slice thickness of 5 mm is required, then in order for a pulse of this bandwidth to excite the slice selectively, there must be a gradient of 1 kHz over a distance of 5 mm. For protons, 1 kHz corresponds to a field of 0.0235 mT, which means that the field gradient must be 0.0235 mT per 5 mm, or 4.7 mT m^{-1}. To achieve a smaller slice thickness, either the bandwidth of the radiofrequency pulse needs to be decreased, or the gradient strength needs to be increased.

Frequency encoding is achieved by collecting the signal in the presence of a read gradient. Typically, the decay may be acquired in the form of 256 consecutive data points over a period of 10 ms. It can be shown that the acquisition time of 10 ms gives rise to a pixel separation of $(10\,\text{ms})^{-1} = 100$ Hz (see Section 7.6.1), and therefore the 256 data points in the resulting image will span a total frequency range of 256×100 Hz $= 25.6$ kHz. If the required field of view is, say, 40 cm, then 40 cm must correspond to a frequency spread of 25.6 kHz, and so the gradient strength must be 25.6 kHz per 40 cm, or 1.5 mT m^{-1}. If the required field of view were to be halved to 20 cm, it may be seen that this could be achieved by doubling the acquisition time or by doubling the field gradient strength. Commonly, it will be desirable to keep the timing of the pulse sequence constant (and moreover the required speed of data collection may preclude an increase in the acquisition time), and so a change in the field of view is generally accomplished by altering the gradient strength.

The process of phase encoding can be considered in an analogous manner. Thus it can be shown that if the phase-encoding gradients are applied for 10 ms, and if 256 phase-encoding steps are to be used, the range of gradient strengths corresponding to a 40 cm field of view would again be 1.5 mT m^{-1}. This range can be achieved by increasing the gradients in the 256 phase-encoding steps from -0.75 mT m^{-1} to $+0.75$ mT m^{-1}. Normally, however, the gradients will be applied for less than 10 ms, say for 4 ms, in which case the required gradients would need to be increased by a factor of $10/4 = 2.5$; i.e. they would need to be stepped from about -1.9 mT m^{-1} to $+1.9$ mT m^{-1}. A reduction in the field of view would again be accomplished by means of a corresponding increase in the gradient strength.

In view of the above calculations, it is not surprising to find that many clinical magnetic resonance instruments employ gradient systems that can achieve strengths of 10 mT m^{-1}, and that even stronger gradients are available with the smaller bore systems used for animal studies. The achievement of rapidly switched gradients of such strength imposes considerable demands upon the power supplies to the gradient coils. Moreover, the rapid switching induces eddy currents in the magnet system which can degrade machine performance. There can be particularly profound consequences for spectroscopy studies, as the eddy currents can severely perturb field homogeneity. These adverse effects can be compensated for, at least in part, by a procedure termed pre-emphasis, which involves changing the time-profile of the gradient pulses in such a way as to cancel out the eddy-current effects. This procedure is not undertaken for each study, but it may have to be carried out in the initial setting up of any newly written pulse sequence, and may need to be checked periodically to correct for any drifts in the nature of the eddy currents.

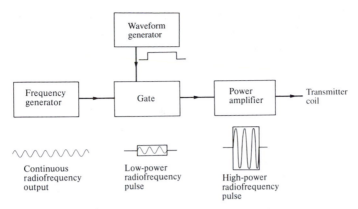

Fig. 7.3 A simple block diagram of the transmitter. Beneath each stage is shown the type of output that is generated. For simplicity, the radiofrequency pulse illustrated here is rectangular in shape.

A more powerful approach to minimizing eddy-current effects is to use active shielding (see Turner 1993 for review). This involves the passage of currents within a screen around the gradient coils, in such a way that the gradient fields are confined within the gradient set. In view of the increasing demand for rapid imaging techniques and for localized ^{1}H MRS of the brain, there is now considerable interest in the use of active shielding, which is becoming much more widely available.

Techniques such as diffusion-weighted imaging benefit from the use of even stronger field gradients than those used for conventional imaging sequences, and for such investigations there is scope for building specially designed gradient sets, for example using coils of a smaller diameter dedicated to studies of the brain. More generally, as a result of the tendency towards increasingly rapid imaging sequences, the new generation of instruments includes systems with gradient strengths of up to $25 \, \text{mT m}^{-1}$.

7.3 THE TRANSMITTER

7.3.1 Introduction

The transmitter generates radiofrequency pulses of the appropriate frequency, power, shape, and timing. It contains a frequency generator, a waveform generator to shape the pulses as required, a 'gate' which switches the transmission on and off at the required times, and a power amplifier which boosts the radiofrequency power to the values that are required in Fourier-transform NMR (see Fig. 7.3). This is all computer controlled by a part of the instrument that is commonly known as the pulse programmer.

As we have seen, the nature of images or spectra is profoundly influenced by the particular pulse sequences that are employed for data acquisition, and a number of the more commonly used sequences are described in some detail in Chapter 8. Here, we discuss a few of the general principles underlying the setting up of radiofrequency pulse sequences.

Let us first consider a simple MRS experiment in which a spectrum is to be accumulated over a given period of time by applying radiofrequency pulses at regular intervals, which we shall call the repetition time TR. The selection of the irradiating frequency is usually fairly straightforward, and it will commonly be within the frequency band of the signals of interest. For the time being, we shall assume that the pulse shape is rectangular, and that the pulses are sufficiently short to excite the whole frequency range uniformly. The main variables that must be considered when setting up the transmitter for this simple experiment are the pulse angle and pulse interval. Considerable care has to be taken in selecting the values of these variables, because any errors can result in substantial losses in the signal-to-noise ratios. Here, we consider how to choose the pulse angles and intervals that provide optimal signal-to-noise ratios.

7.3.2 Accumulation of signals — T_1 and the effects of saturation

For most spectroscopy studies, the signals obtained on Fourier transformation of a single free induction decay are too weak to be clearly distinguishable from noise. It is therefore almost always necessary to improve the signal-to-noise ratio by accumulating a large number of FIDs and Fourier transforming the sum of these FIDs. The accumulation of N FIDs leads to an improvement of $\sqrt{N}$ in the signal-to-noise ratio, because the signal increases by a factor of N whereas the noise, being random, increases only by $\sqrt{N}$.

The choice of time interval between consecutive radiofrequency pulses is influenced by the process of saturation, i.e. by the reduction in signal intensity that is associated with incomplete T_1 relaxation. Consider, for example, the effects of a series of 90° radiofrequency pulses. Each of these pulses essentially 'samples' the z-component of magnetization M_z that exists immediately prior to the pulse. It does this by tilting the magnetization away from the z-axis into the xy-plane, thereby generating xy-magnetization, M_{xy}, that produces the observed signal. The magnetization M_{xy} decays with a time constant T_2^*, and so almost disappears after a time of about $4T_2^*$. It might therefore seem sensible to apply consecutive pulses at intervals of $4T_2^*$, so that signal can be observed more or less continuously. However, for biological samples the time constant T_1 that characterizes the recovery of M_z is usually very much longer than T_2^*. As a result, M_z (which is zero immediately after being tilted into the xy-plane) may have recovered very little after the time $4T_2^*$, in which case it is certainly not optimal to apply each 90° pulse immediately following the previous FID, for correspondingly little signal would be generated.

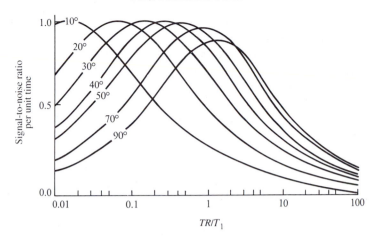

Fig. 7.4 The signal-to-noise ratio per unit time expressed as a function of the interval between successive radiofrequency pulses, for various pulse angles. (From Shaw 1976.)

Conversely, it is time-wasting to wait until M_z has almost completely recovered to its equilibrium value M_0. Thus the setting of the pulse interval involves a compromise. If the interval is too small, there is a danger of saturating the signals, with a consequent loss in signal because the magnetization prior to each pulse is much less than M_0. If the interval is too large, then in a given period of time too few scans are collected, and since the signal-to-noise ratio increases according to the square root of the number of scans this also adversely affects the signal-to-noise ratio.

It is not unduly difficult to calculate how the signal-to-noise ratio obtained in a given time varies with pulse angle and interval; the relevant equation is given by Becker *et al.* (1979) following the analysis of Ernst and Anderson (1966), and the results are shown in Fig. 7.4. From this diagram, it can be seen that if we wish to apply 90° pulses then the optimal signal-to-noise ratio is obtained if the pulses are applied at intervals of 1.25 T_1, where T_1 is the spin–lattice relaxation time of the signal of interest. However, it is even better to apply shorter pulses rather more rapidly. For example, 45° pulses every 0.35 T_1 will generate a 10 per cent higher signal-to-noise ratio in a given period of time than 90° pulses applied every 1.25 T_1. The optimum angle for a given pulse interval is known as the Ernst angle, α; it is given by the expression

$$\cos \alpha = \exp\left(-TR/T_1\right) \tag{7.1}$$

and is plotted in Fig. 7.5.

The curve plotted in Fig. 7.4 suggests that the optimal approach might be to use very small pulse intervals. However, in practice there will be certain

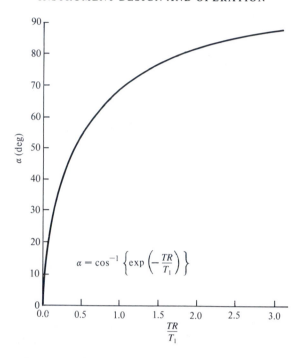

$$\alpha = \cos^{-1}\left\{\exp\left(-\frac{TR}{T_1}\right)\right\}$$

Fig. 7.5 The Ernst angle as a function of pulse interval. (From Shaw 1976.)

lower limits to the choice of interval, which will be determined by other features of the pulse sequence and by the T_2 (or T_2^*) values of the signals. It may be seen from Fig. 7.4 that a safe choice, which would suit many situations, would be to apply 45° pulses every 0.35 T_1. For imaging studies, these dependencies are perhaps of most practical benefit when setting up rapid imaging sequences such as FLASH, which utilize low-angle excitation pulses applied at time intervals that may be two orders of magnitude less than T_1.

7.3.3 Selection of pulse parameters

Figures 7.4 and 7.5 provide a theoretical basis for selecting the appropriate pulse angle and interval. However, in order to put this into practice, it is first necessary to establish the transmitter settings that generate the required pulse angles, and also to know the T_1 values of the signals of interest, as the relationships of Figs 7.4 and 7.5 are expressed in terms of the ratio TR/T_1. We shall first consider the setting of pulse angles.

We noted in Section 5.3.1 that a radiofrequency pulse of duration or width t_p tilts the magnetization through an angle θ given by

$$\theta = \gamma B_1 t_p \tag{7.2}$$

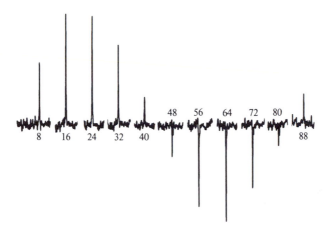

Fig. 7.6 Signal amplitude from a single scan, plotted as a function of pulse width. The pulse widths corresponding to each signal are given in µs.

This equation shows that θ, the pulse angle, is linearly related to the strength of the irradiating B_1 field and to the pulse width. Therefore the pulse angle may be altered by changing either the B_1 field or the pulse width. Traditionally, the width of the pulses is changed for conventional spectroscopy studies, while for a variety of practical reasons the strength of the B_1 field is normally altered for imaging and localized spectroscopy. (Note that the irradiating power is proportional to B_1^2, and so a doubling of B_1 requires a four-fold increase in power.) To set a particular pulse angle, it is first necessary to define the settings that generate a 90° or 180° pulse; the pulse power or width can then be adjusted in order to generate the required angle. The definition of these settings is based on a calibration procedure of the type shown in Fig. 7.6, which shows the intensity of a signal plotted as a function of pulse width; it is apparent that in this illustrative example pulse widths of about 20 and 42 µs correspond to 90° and 180° pulses, respectively. Therefore 45° pulses would be obtained by setting the pulse width to 10 µs. For many clinical systems, such calibration procedures are automated, so that it is simply necessary to enter the appropriate angle into the list of sequence parameters. However, automated procedures of this type are not always available, in which case it will be necessary to carry out a calibration as in Fig. 7.6. This is not necessarily straightforward; for example, problems can arise if the B_1 field is not uniform. Problems also arise if none of the signals is strong enough to provide adequate signal-to-noise ratios in a single scan. Under these circumstances, the calibration procedure may prove to be very time-consuming.

Fortunately, for the study of small specimens, it is quite likely that the 90° pulse settings for a given radiofrequency coil will remain fairly constant from one experiment to another; this may be known on the basis of previous experience. However, this is not always the case, primarily because the performance of the transmitter coil is dependent upon a number of factors, including the nature of the sample, in particular its size and electrical conductivity (see Sections 7.4.4 and 7.4.6). If there is some uncertainty about the settings, then they should be checked, particularly if there is a special need for accuracy, for example when carrying out T_1 or T_2 measurements.

In order to use the plots of Figs 7.4 and 7.5 to optimize pulsing conditions, it is also necessary to know the approximate T_1 values of the resonances. If these are not known on the basis of prior studies, then they may be determined using one of the pulse sequences described in Section 8.2. A slightly different approach to optimizing the pulsing conditions, which is rather more empirical and less time-consuming, is to decide to apply pulses of a given angle, say 45°, and to determine by trial and error the optimum interval for this particular angle.

Of course, the choice of pulsing parameters becomes somewhat more complicated if different signals of interest have differing T_1 values, because optimizing the signal-to-noise ratio for one signal will certainly not optimize the signal-to-noise ratio for another. Therefore, in spectroscopy a certain amount of compromise is often necessary when selecting pulsing conditions. One of the difficulties is that, since signals with different T_1 values will undergo different amounts of saturation, the relative peak areas will not necessarily reflect the relative concentrations of the corresponding metabolites. These effects have to be taken into account when interpreting spectra; if information is required about metabolite levels it will be necessary to know the extent to which the various spectral lines are saturated. If the extent of saturation is not known from previous experience, then some studies will have to be undertaken under non-saturating conditions, which is necessarily time-consuming. The term 'saturation factor' is sometimes used to express the degree of saturation of a signal; it is equal to the area of the signal measured under non-saturating conditions divided by its area measured under the particular pulsing conditions of the study.

7.3.4 Contrast-to-noise ratios

While the above discussion has been concerned with spectroscopy, similar arguments hold for imaging studies, as most imaging sequences also rely on the acquisition of many scans acquired at intervals of TR; for example, a standard two-dimensional imaging sequence may involve the consecutive acquisition of, say, 128 or 256 phase-encoding steps. In imaging, however, differential saturation effects associated with differences in T_1 are not regarded as a nuisance, as in spectroscopy; instead they are exploited in order to generate contrast. Therefore, the parameter of interest is commonly not the signal-to-noise ratio, but rather the signal difference, or contrast-to-noise ratio. If contrast-to-noise is the important criterion, then Figs 7.4 and 7.5 can again be exploited, but in a somewhat different way, for the emphasis will be on using the various curves to optimize signal *differences* between tissues with differing T_1 values. More generally, contrast in imaging may be dependent on a variety of parameters, including T_1, T_2, T_2^*, and flow rates, and more complex calculations will be needed to determine the pulsing conditions that best highlight specific tissues or pathologies.

7.3.5 The effective bandwidth of radiofrequency pulses

When considering the effects of a radiofrequency pulse, we have assumed so far that its frequency is equal to the resonance frequency of the nuclei. However, this assumption cannot be valid for all of the signals occupying a range of frequencies. We must therefore consider how effective a pulse is when its frequency (which we now define as ν_1) differs from the resonance frequency ν_0 of the nuclei.

In Section 5.6.1 we briefly considered the effects of finite pulse width. In particular, it was noted that, as the pulse width increases, we might expect the effective bandwidth of the pulse to decrease. However, it was also pointed out that the precise frequency *response* to a radiofrequency pulse differs slightly from the frequency distribution of the pulse itself.[1]

In order to determine the frequency response, we must resort to calculations using Bloch equations of the type described by Meakin and Jesson (1973). Some results from these calculations are illustrated in Fig. 7.7. Six simulated sets of data are shown, each of which represents the Fourier transform of the response to a single rectangular 90° pulse. Each spectrum corresponds to a different B_1 field strength and hence to a different value for the 90° pulse length, t_p^{90}; as we have seen in Section 5.3.1, B_1 and t_p^{90} are related to each other through the equation

$$\gamma B_1 t_p^{90} = \pi/2 \qquad (7.3)$$

The horizontal axis in Fig. 7.7 represents the difference between the resonance frequency and the frequency of the applied B_1 field; this difference, $\nu_0 - \nu_1$, is commonly called the offset. It can be seen that, as the offset increases, the intensity of the corresponding signal declines and its phase changes. The effects become more marked as t_p^{90} increases, i.e. as B_1 decreases. Theory shows that if intensity errors of less than 2 per cent are required, the value of t_p^{90} must be less than $1/4\Delta$ where Δ is the maximum offset; this corresponds to a value of t_p^{90} of less than 50 μs for an offset of 5 kHz.

Figure 7.7 refers to spectra obtained by application of 90° pulses. The

---

[1] There would be no difference between the two if NMR were a linear system, i.e. if the output from the system were linearly proportional to the input. This is certainly not the case for NMR, because the response to a 180° pulse is *not* double the response to a 90° pulse.

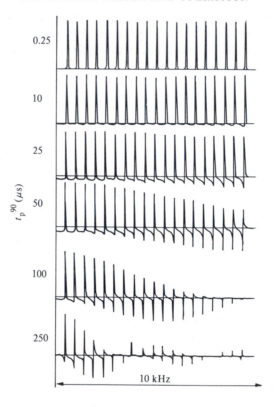

Fig. 7.7 The effect of finite pulse power (i.e. length of a 90° pulse) on signal phase and amplitude, plotted as a function of offset from 0 to 10 kHz. (From Meakin and Jesson 1973.)

intensity variations in the two lowest plots of Fig. 7.7 are clearly unacceptable, but the situation can be improved considerably by applying smaller angle pulses. For example, a 45° pulse of length 50 µs has about twice the effective bandwidth of a 90° pulse of length 100 µs, and a 22° pulse of length 25 µs has a bandwidth about four times as great, i.e. there is an approximate inverse relationship between the duration of a pulse and its effective bandwidth, as might be expected from the arguments in Section 5.7.1. Therefore, if the weakness of the B_1 field could conceivably lead to bandwidth problems, it would be preferable to apply small-angle pulses at intervals that can be determined by reference to eqn (7.1).

7.3.6 The shape of radiofrequency pulses

The results shown above were derived for radiofrequency pulses with a rectangular profile. Such pulses are commonly used when no frequency selectivity is required. However, when pulses are required to be frequency selective, as in slice selection, then a rectangular profile is no longer optimal. A sinc-shaped pulse provides much better slice selectivity, because it generates a frequency response that is much more rectangular in shape than that produced by a rectangular pulse. In modern NMR instruments, the radio-frequency pulse is represented digitally, typically by 128–512 data points, and can, at least in principle, be programmed to take on any given shape. There has been a great deal of progress in the development of new types of pulse that give enhanced performance including, for example, better slice selectivity, improved solvent suppression, or more complete excitation or inversion of the nuclear spins (see van Zijl and Moonen 1992; Garwood and Ugurbil 1992; Morris 1992). With increasingly powerful computation for assessing the effects of complex pulses, many further developments can be envisaged.

The terms 'hard' and 'soft' are often used to describe radiofrequency pulses, 'hard' pulses being short, rectangular, and non-selective, while 'soft' pulses are longer, shaped, and frequency selective.

7.3.7 Double irradiation

In addition to the radiofrequency pulses that are used for routine data accumulation, it is sometimes useful, or even necessary, to apply radiation at a second frequency, or range of frequencies. Some of the pulse sequences described in Chapter 8, including those used for magnetization transfer and for spin decoupling, require this type of double irradiation. The ease or feasibility of carrying out such studies depends on the nature of the instrument, research-oriented machines commonly having greater versatility and flexibility in this respect.

7.4 THE RADIOFREQUENCY COIL(S)

Radiofrequency coils are used for transmitting the B_1 field into the region of interest, and for detecting the resulting signal. In some cases, the same coil is used for transmission and reception, while in others it may be preferable to use separate transmit and receive coils. The versatility and performance of NMR instruments are strongly dependent upon the characteristics of the radiofrequency coils, and in this section we discuss some of the general features and requirements of these coils. We begin by discussing transmission.

7.4.1 Requirements for transmission

The radiofrequency coil needs to generate 90° or 180° pulses of sufficient power to give uniform excitation over the required band of frequencies. Figure 7.7 illustrates the problems that may arise if the 90° pulses are too long (i.e. if the radiofrequency field B_1 is too weak); signals that are too far off-resonance suffer intensity reductions and phase changes. A convenient and effective way of achieving adequate B_1 fields is to incorporate the coil into a tuned circuit that resonates at the required irradiation frequency (see Section 7.4.6). At this frequency, there is an amplification of the electric current that passes through the coil, as a result of which a large B_1 field can be generated.

The second important consideration is B_1 homogeneity. If the B_1 field is not uniform over the region of interest, then nuclei in different parts of the region will experience different B_1 fields and as a result the pulse angle will vary with spatial location. This may in turn lead to unacceptable variations in signal intensities. Radiofrequency coils vary in the degree of B_1 homogeneity that they provide, and while increasingly sophisticated methods are becoming available for overcoming the consequences of B_1 inhomogeneity (see Garwood and Ugurbil 1992), B_1 homogeneity remains an important criterion in the design of radiofrequency coils.

7.4.2 Requirements for signal reception

The requirements for signal reception are in many respects consistent with those for transmission. First, there is a need to maximize the signal-to-noise ratio. In practice, this means that there must be optimal magnetic coupling of the coil to the sample, and it is not surprising to find that a coil that efficently generates a strong B_1 field also has good characteristics for signal reception. Secondly, it is important to ensure that the coil is equally effective at picking up signal from all the regions of interest. A coil with good B_1 homogeneity fulfils this requirement. In practice, therefore, we find that the requirements for transmission and reception are often (although not always) completely compatible with each other, which explains why it is common practice to use the same coil for transmission and reception. However, in some circumstances it remains preferable to use separate transmit and receive coils.

7.4.3 Types of radiofrequency coil

We have seen that, for both transmission and reception, it is important for the coil to produce a strong, homogeneous B_1 field. Moreover, this B_1 field must be oriented in the plane perpendicular to B_0; any components of the B_1 field that are parallel to B_0 will be ineffective. Several types of coil geometry have acceptable characteristics; three of these are shown in Fig. 7.8. Of the three, the solenoidal and saddle-shaped coils are the most traditional. While solenoidal coils tend to have better performance than saddle-shaped coils in terms of signal-to-noise ratio, they produce a field along the axis, and this generates practical difficulties if they are to be used in conjunction with conventional superconducting magnets; for B_1 to be oriented perpendicular to B_0, the axis of the radiofrequency coil must be perpendicular to the axis

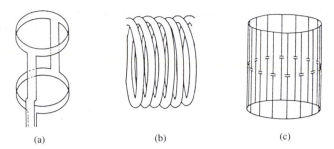

Fig. 7.8 Three types of coil design: (a) a saddle-shaped coil, (b) a solenoidal coil, and (c) a bird-cage coil.

of the magnet, and this is simply impractical for most investigations because the coil impedes entry of the sample. Solenoidal coils can, however, be used in conjunction with other types of magnet system (for example with permanent magnets), and their superior performance in terms of signal-to-noise ratio helps to offset the relatively poor sensitivity associated with low-field permanent or resistive systems.

Saddle-shaped coils are extensively used with superconducting systems, both in very high field, narrow bore magnets, and for clinical studies. They are relatively simple to manufacture, and can be readily shaped, for example to surround a glass sample tube or the human body. However, for clinical studies, increasing use is being made of an alternative design, termed a birdcage coil (Hayes *et al.* 1985; Fig. 7.8(c)). This design of coil is a relatively new development, and can provide improved B_1 homogeneity and sensitivity.

Finally, considerable use has been made of a type of coil known as a surface coil (see Bosch and Ackerman 1992). In its simplest form, this consists of a circular loop of wire which is placed adjacent to the region of interest, which may, for example, be a limb muscle or in the spine. Such coils provide a particularly sensitive means of detecting signal from superficial regions (where their B_1 fields are strongest), but this is achieved at the expense of B_1 homogeneity. Coils of this type were used for early studies of blood flow (see Morse and Singer 1970), and provided the original means of obtaining spectra from localized regions of the body non-invasively (Ackerman *et al.* 1980). Surface coils continue to be used extensively in both spectroscopy and imaging studies (see Sections 7.4.5 and 7.4.6). However, as a result of their poor B_1 homogeneity, they tend to be used primarily for reception of signal, in conjunction with a larger transmitter coil that provides better B_1 homogeneity and can hence generate uniform radiofrequency pulses throughout the region of interest.

Numerous designs have been proposed for radiofrequency coils (see Link 1992 for review), and this remains an area of active research for, as discussed below, improvements in coil design can have a strong impact on machine performance, in particular on the signal-to-noise ratios that can be achieved.

7.4.4 Theory underlying coil design and performance

The theory underlying coil design and performance is best understood in terms of the approach developed by Hoult and Richards (1976), and here we outline briefly some of their major findings.

Let us consider a solenoidal coil, as shown in Fig. 7.8(b), surrounding a sample. Suppose that a 90° pulse has been applied by passage of a suitable current through the coil, and suppose that the same coil is to be used for detection of the resulting signal. The xy-magnetization that is generated by the pulse precesses with angular frequency ω_0, and induces an electromotive force (e.m.f.) of the same frequency in the coil. The magnitude E of this e.m.f. is given by Faraday's law of electromagnetic induction:

$$E = -\frac{d\Phi}{dt} \tag{7.4}$$

where Φ is the magnetic flux linkage through the coil. The e.m.f., E, constitutes the signal that, after much processing, appears in the final image or spectrum. Now it is reasonable to expect that the magnetic flux passing through the coil, and hence the induced e.m.f., would decrease if the coil were moved away from the sample; we could say that under these conditions the coupling between sample and coil would decrease. Similarly, as the coupling decreases, the B_1 field that would be produced at the sample if *unit* current were passed through the coil would also decrease. It turns out that there is a well-defined relationship between the magnitude of the induced e.m.f. and the strength of the B_1 field generated in the sample by unit current in the coil. This relationship was used by Hoult and Richards to obtain an expression for the induced e.m.f.; for a region of volume V_s in a homogeneous B_1 field, the induced e.m.f. following a 90° pulse is

$$E = \omega_0 B_{1(u)} M_0 V_s \tag{7.5}$$

where $B_{1(u)}$ is the field generated in the xy-plane by unit current passing through the coil. The nuclear magnetization M_0 is given by

$$M_0 = C\gamma^2\hbar^2 I(I + 1)B_0/3kT_s \tag{7.6}$$

where C is the number of resonant nuclei per unit volume, γ is the gyromagnetic ratio of these nuclei, T_s is the temperature, and k is the Boltzmann constant.

Equations (7.5) and (7.6) enable the signal generated in the coil to be evaluated in terms of parameters that are either known or can be estimated. In order to evaluate the signal-to-noise ratio, it is now necessary to obtain an expression for the noise that is generated. One component of the noise originates in the coil itself, and consists of random fluctuations of e.m.f. associated with the Brownian motion of the electrons within the coil. It can be shown that the root mean square noise e.m.f. over a bandwidth Δv, at the resonant frequency is given by

$$N = (4kT_c R_c \Delta v)^{1/2} \qquad (7.7)$$

where R_c is the resistance of the coil and T_c is its temperature.

In a correctly designed NMR system, the only other significant source of noise should be that generated by the sample. Living systems have fairly high concentrations of ions, and as a result they conduct electricity. The electrical conductivity of tissue is approximately equal to that of a solution of 100 mM NaCl, and for large samples this can have profound consequences, first with respect to the generation of noise, and secondly with respect to the penetration depth of radiofrequency fields into the sample; the radiofrequency field B_1 has a limited depth of penetration into conducting samples, the depth decreasing according to $\sqrt{v_0}$. In principle, this manifestation of the 'skin effect' imposes an upper limit on the sample size that can generate useful signals, but in practice there do not appear to be major problems at the field strengths that are most commonly used for either clinical or basic studies. The generation of noise is a more important issue, which we now consider.

Following a radiofrequency pulse, we have seen that the magnetization developed in the xy-plane induces an e.m.f. in the receiving coil. A back e.m.f. is also induced in the sample itself, and if the sample is conducting, currents flow within it and these so-called eddy currents dissipate power. This power dissipation is analogous to power losses within the resistance of the coil itself, and therefore represents an additional source of resistance R_m. The subscript 'm' refers to the fact that the resistance is associated with the magnetic field of the radiofrequency coil, and the expression 'magnetic losses' can be used to describe the power dissipation. The total effective resistance R_t is then equal to $R_c + R_m$, where R_c is the inherent resistance of the coil.

It can be shown that, for a cylindrical sample enclosed by a solenoidal coil,

$$R_m = B_{1(u)}^2 V_s^2 \omega_0^2 \sigma / 16\pi g \qquad (7.8)$$

where σ is the conductivity of the sample and $2g$ is the length of both sample and coil. Therefore, by combining eqns (7.5) to (7.8) we can obtain the following expression for the signal-to-noise ratio:

$$\Psi \propto C\omega_0 B_{1(u)} V_s B_0 / \{R_c + [B_{1(u)}^2 V_s^2 \omega_0^2 \sigma / 16\pi g]\}^{1/2} \qquad (7.9)$$

7.4.5 The signal-to-noise ratio in NMR

While the expression given in eqn (7.9) was derived for a specific coil and sample geometry, it can be used to illustrate more generally the important factors influencing signal-to-noise ratios in imaging and spectroscopy studies. In particular:

1. The signal-to-noise ratio is proportional to the volume of each volume element. Thus, if we wish to reduce the size of each voxel from, say, $2 \times 2 \times 2\,mm^3$ to $1 \times 1 \times 1\,mm^3$ (without changing the radiofrequency coil), we will lose a factor of 8 in signal-to-noise. In order to recover the signal-to-noise of the larger voxel, it would be necessary to increase the number of of acquisitions by a factor of $8^2 = 64$. Therefore, a relatively modest improvement in linear spatial resolution can have very severe consequences in terms of signal-to-noise or data acquisition time.

2. The signal-to-noise ratio is proportional to the concentration of resonant nuclei; hence water (at a mean concentration of about 40 M in tissue) provides the basis for imaging, while at the other end of the concentration range, those metabolites that are present *in vivo* at concentrations below about $100\,\mu M$ will generally not give rise to detectable signals.

3. For the examination of superficial regions, the use of surface coils is desirable, firstly because their relatively small size tends to improve their intrinsic detection characteristics, and also because sample noise is generated only by the relatively small volume 'seen' by the coil. More generally, signal-to-noise ratios tend to be optimized by the use of receiver coils that are fitted reasonably closely around, or adjacent to, the region of interest; coils that are unnecessarily large will give relatively poor sensitivity. However, it should be noted that if the coil is placed too near the sample, then electric lines of force passing through the sample can generate additional resistive losses, resulting in a decrease in the signal-to-noise ratio (Hoult and Lauterbur 1979; Gadian and Robinson 1979). These 'dielectric losses' should be avoided, if at all possible.

4. The dependence of signal-to-noise ratios on field and frequency is somewhat complex, and depends on the relative amounts of noise generated by the coil and by the sample. However, it may be shown from eqn (7.9) that if the coil noise dominates, then the signal-to-noise ratio for any given nucleus will increase approximately according to $B_0^{7/4}$; if the sample noise dominates, the increase will be according to B_0. The latter dependence will generally hold at high field strengths when using large coils and samples.

7.4.6 Practical aspects of coil design and circuitry

Power is transferred to and from the radiofrequency coil by means of a coaxial cable, and efficient transfer only takes place if the electrical characteristics of the coil circuitry match those of the cable. Cables usually have a 'characteristic impedance' of 50 ohms, and it is therefore essential for the coil circuit to appear to behave like a 50-ohm resistance. The characteristics of the coil

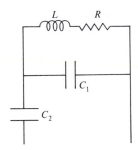

Fig. 7.9 A simple tuned circuit.

itself are very different from this, and they must therefore be transformed in some way to the required 50 ohms. This can be conveniently and effectively achieved using a tuned circuit of the type shown in Fig. 7.9. The coil is represented as having both inductance (L) and resistance (R), and is connected to a tuning (C_1) and matching (C_2) capacitor. The tuning capacitor and the coil constitute a tuned circuit which has a resonance frequency ω_0 given by

$$\omega_0^2 = 1/LC \qquad\qquad (7.10)$$

At this frequency, there is an amplification of current through the coil enabling a large magnetic field B_1 to be generated for a given input power. However, the tuning capacitor in itself does not ensure efficient power transfer, because the circuit will still not appear to behave as a 50-ohm resistance. What is required is an additional 'matching' capacitor C_2. It is not intuitively obvious why a circuit such as this can appear equivalent to a 50-ohm resistance, but nevertheless it works; the logic behind the circuit has been discussed by Hoult (1978).

The performance or quality of a tuned circuit is commonly expressed in terms of the 'quality factor' Q, which is given by the expression

$$Q = \omega_0 L/R \qquad\qquad (7.11)$$

A coil with a high Q-value has a correspondingly sharp frequency response at the resonance frequency, and will tend to have high performance characteristics in terms of the signal-to-noise ratio that it can provide.

Prior to starting any NMR study, it is essential to ensure that the coil is tuned and matched, i.e. to ensure that the values of C_1 and C_2 are such that the circuit behaves like a resistance of 50 ohms at the resonance frequency. This procedure is commonly automated; if not, a simple iterative procedure involving adjustments of the two capacitors needs to be carried out manually. One of the reasons why adjustments may be required from one study to another is that the loading or effective resistance generated by the sample may vary, depending, for example, on the size of the sample under investigation. It should be noted that if the tuning and matching capacitors need to be adjusted between studies, this reflects a change in the power characteristics of the radiofrequency coil, and as a result the settings that are required to generate 90° or 180° pulse angles (see Section 7.3.3) will also change.

A number of general comments can be made about the choice of coil design. For studies of solutions, including body fluids and tissue extracts, it is common practice to use the standard manufacturers' coils, which form an intrinsic part of the commercial probes that are bought as part of the complete system. These coils are designed not just for optimal sensitivity and B_1 homogeneity, but also to introduce as little perturbation of B_0 homogeneity as possible; we have seen that B_0 homogeneity needs to be particularly good for high-field solution studies. Such coils and probes may also be suitable for analyses of preparations such as perfused cells.

For animal studies, more flexibility and versatility may be desirable; for example surface coils of various sizes may be needed. Provided that appropriate expertise is available, such coil systems can be built in-house; alternatively the manufacturers may be able to provide the required coils. The circuitry will generally be somewhat more sophisticated than that illustrated in Fig. 7.9, particularly if the coils need to be tuned to more than one frequency (Schnall 1992), but this remains a part of the NMR system that is most readily accessible to modification and development by the well-qualified user.

For clinical studies, it is relatively uncommon for users to design their own coils. This is partly for reasons of safety, and partly because of continued developments by the manufacturers, who are providing an increasing range of user-friendly coils. For example, in addition to standard head and body coils, surface coils are available, e.g. as receivers for imaging the spine, or in the form of flexible coils conformed to the contours of the body, (for the shoulder, neck, or jaw, for example). In addition, coils can be constructed in such a way that they generate a rotating B_1 field. The main advantage of such 'quadrature' coils is that they produce an increase in signal-to-noise ratio by a factor of up to $\sqrt{2}$. This improvement is achieved because such coils effectively detect signal components along both the x- and y-directions, rather than along just one of these directions.

7.5 THE RECEIVER

7.5.1 The basic design

A simplified block diagram of an NMR receiver is shown in Fig. 7.10. This diagram serves to illustrate a number of the important principles of signal reception.

The signal and noise emerge from the receiver coil in the form of a very small voltage, which may be as low as nanovolts. It is therefore essential to ensure that in the early stages of amplification very little additional noise

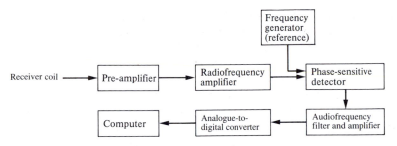

Fig. 7.10 A simplified block diagram of the spectrometer receiver.

is introduced, because only a small noise level superimposed upon a few nanovolts could greatly degrade the signal-to-noise ratio that is finally obtained. In this respect, the pre-amplifier in particular must be designed and built with great care.

Following the initial amplification steps, the signals enter the phase-sensitive detectors. Before discussing the function of these devices, it is important to consider the nature of the image or spectrum that is eventually to be obtained. Typically, the signals that constitute a spectrum or image may have a frequency of, say, 63 MHz, occupying a frequency range of, say, a few kilohertz. However, the computer is not capable of detecting and distinguishing between signals at such high frequencies, nor is the filter capable of selecting a region occupying only a few kilohertz at a frequency of 63 MHz. Therefore, the approach taken is to subtract the frequency, v_1, of the applied B_1 field from the frequency, v_0, of the signals, thereby generating a group of frequencies in the range zero to a few kilohertz. These frequencies, which are in the audiofrequency range, can be handled by the computer and can be filtered adequately. The required subtraction of frequencies is accomplished with the aid of a device known as a phase-sensitive detector. This device combines the signal with a reference waveform which in the simplest case oscillates at the frequency v_1 of the field B_1. Mathematical analysis shows that there is an output from the phase-sensitive detector of the form $\cos \{2\pi (v_0 - v_1) t + \phi\}$. This represents a signal of frequency $v_0 - v_1$ which is equal to the difference in frequency between the signal and the reference. The phase angle ϕ is equal to the phase difference between the signal and the reference; hence the name of the device.

The audiofrequency output from the phase-sensitive detector is filtered and then amplified to the required level before being fed, via the analogue-to-digital converter (ADC), into the computer. The filter removes high-frequency noise, which if present would have adverse effects on the signal-to-noise ratio (see Section 7.6.1).

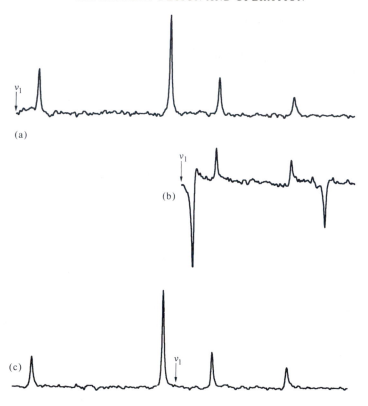

Fig. 7.11 The positioning of the reference frequency. (a) The reference is to one side of the signals, and the signals are displayed correctly. (b) The reference is positioned among the signals, and this causes problems when not using quadrature detection; see Section 7.5.2. (c) When using quadrature detection the reference can be placed among the signals.

7.5.2 The use of quadrature detection

An important drawback of the detection system considered above is that information is lost as to whether $v_0 - v_1$ is positive or negative. Therefore, following phase-sensitive detection and Fourier transformation, all signals will appear to one side of the reference frequency, regardless of the sign of $v_0 - v_1$. This effect is illustrated in Fig. 7.11. Figure 7.11(a) shows a spectrum containing four signals detected in the manner described above with the reference frequency positioned to one side of the lines. Figure 7.11(b) shows the effect of placing the reference among the signals. Two of the signals are positioned correctly but the positions of the other two have been reflected about the reference frequency. For exactly the same reasons, noise is reflected about the reference frequency. Therefore two sets of noise contribute to the

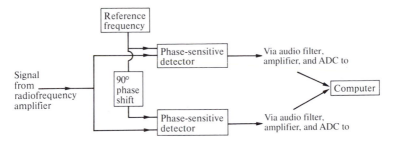

Fig. 7.12 Simplified block diagram for a receiver employing quadrature detection.

spectrum instead of one. Since noise is random, this causes the noise amplitude to increase by a factor of $\sqrt{2}$ and the signal-to-noise ratio to decrease by a factor of $\sqrt{2}$.

Clearly, it would be advantageous to employ a method that is capable of overcoming this problem. The use of two phase-sensitive detectors in quadrature provides such a method. Signal and reference are fed into two separate phase-sensitive detectors, and as illustrated in Fig. 7.12 the phase of the reference is shifted by 90° in one of the detectors. It is this 90° phase shift to which the term 'in quadrature' refers. The outputs from the two phase-sensitive detectors differ in phase by 90°, but are similar in all other respects. They are both filtered and amplified before being fed into two separate channels of the computer.

Figure 7.13 shows FIDs observed in the two channels (a) when $v_0 - v_1 < 0$ and (b) when $v_0 - v_1 > 0$. The signals in channel 1 are the same in the two cases but those in channel 2 are not; when $v_0 - v_1 > 0$ the phase of the signal in channel 2 leads that in channel 1 by 90°, whereas when $v_0 - v_1 < 0$ the phase in channel 2 lags that of channel 1 by 90°. It is this difference that, on Fourier transformation, leads to the separation of positive and negative frequencies.

The effects of this method of separation are illustrated by the spectrum of Fig. 7.11(c) which was obtained under similar conditions to those used for the spectrum of Fig. 7.11(b), except that quadrature detection was used. In contrast to the spectrum of Fig. 7.11(c), no folding over is observed when the reference frequency is in the middle of the spectrum. In fact, spectra obtained with quadrature detection have zero offset at the centre; negative frequencies are displayed to the right and positive frequencies to the left.

Quadrature detection has two main advantages over detection using a single phase-sensitive detector. First, noise is not folded about the reference frequency, and as a result there is an improvement in signal-to-noise ratio by a factor of $\sqrt{2}$ for a given number of scans. This corresponds to a saving in accumulation time of a factor of two, which can be extremely valuable. Secondly, the reference frequency can be placed in the middle of the frequency

(a)

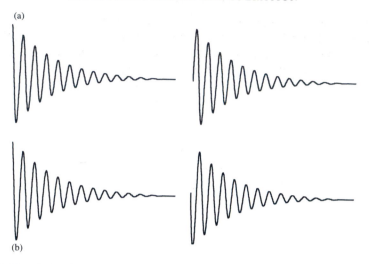

(b)

Fig. 7.13 Free induction decays observed in the two channels when employing quadrature detection. (a) $(\nu_0 - \nu_1) < 0$, and (b) $(\nu_0 - \nu_1) > 0$.

range of interest. As a result, the maximum offset in a spectrum or image covering a width of W kHz is only $W/2$ kHz, as opposed to W kHz when a single-phase-sensitive detector is used. This considerably reduces bandwidth problems that may arise from the effects of finite pulse width (see Section 7.3.5). The ability to place the reference frequency anywhere within the range of frequencies to be detected is also of considerable importance.

A potential difficulty with the use of quadrature detection is that unless the two signals in the computer have exactly equal amplitudes and are exactly 90° out of phase with each other, there will be artefactual signals reflected about zero frequency. For this reason special data-routing and phase-cycling techniques were developed in order to correct for inaccuracies in the gains and phase settings of the receiver channels (Stejskal and Schaeffer 1974; Hoult and Richards 1975; Redfield and Kunz 1975). Quadrature detection, in conjunction, as necessary, with such phase-cycling and data-routing techniques, is now a routine feature of NMR systems.

7.5.3 Adjustment of the receiver settings

The main variables that may require adjustment are the gain (amplification) settings and the filters. The role of the filters is related to the manner in which the data are collected by the computer, and a discussion of filter settings is therefore given in Section 7.6.1. The gain settings should be low enough to ensure that the signal or noise does not overload any of the receiver stages. At the same time, the settings should be sufficiently high to ensure that the receiver itself contributes a negligible amount of noise.

In some cases, there may be unintended 'leakage' of the reference frequency into the signal channel of the receiver. Phase-sensitive detection of this leakage generates an output of zero frequency, i.e. a direct current (d.c.) output. This manifests itself in spectroscopy as an artefactual signal at zero frequency, or in imaging as an artefactual band passing through the centre of the image in the phase-encoding direction. Such effects can be eliminated or at least reduced by simple adjustment of the d.c. level, or more generally by alternating the phase of the radiofrequency excitation by 180° in consecutive acquisitions. With this phase-cycling scheme, the signal is alternately positive and negative, while the leakage does not alternate in phase. Therefore, alternate addition and subtraction of consecutive acquisitions should cause the required signal to add, and the leakage to cancel out.

7.6 THE COMPUTER

The computer has a wide range of functions. Its main functions are: (i) to control the radiofrequency and field gradient pulses; (ii) to accumulate the data; and (iii) to process and display the data. In Sections 7.2 and 7.3 we discussed the gradient and radiofrequency pulsing conditions and in Chapter 8 we describe specific pulse sequences; here we discuss data accumulation, processing, and display.

7.6.1 Collection of data

The signals that emerge from the receiver are continuous, whereas the computer samples the incoming signals at discrete time intervals, and it is particularly important to appreciate the implications of this.

Consider a free induction decay representing a single frequency, as shown in Fig. 7.14(a). The time between successive samplings of the decay is termed the sampling interval, or alternatively the dwell time. It is important to know how frequent the sampling must be in order for the resulting discrete pattern to represent correctly the signal that is being sampled. There is a theorem due to Nyquist that shows that the sampling must be performed at least twice per wave cycle in order for the resulting frequency to be equal to the actual frequency of the wave. This is illustrated in Fig. 7.14. In Fig. 7.14(b) the sampling of the FID shown in Fig. 7.14(a) is sufficiently rapid, whereas in Fig. 7.14(c) it is not; the apparent frequency in the latter case is less than it should be because the frequency of sampling is less than twice the actual frequency of the waveform. Thus the sampling interval determines the maximum frequency observed after computer detection; if the interval is d milliseconds, the maximum frequency is $1000/2d$ Hz. Therefore, if a single-phase-sensitive detector were to be used in the receiver, the frequency width of the resulting spectrum or image would be $1000/2d$ Hz. Using quadrature

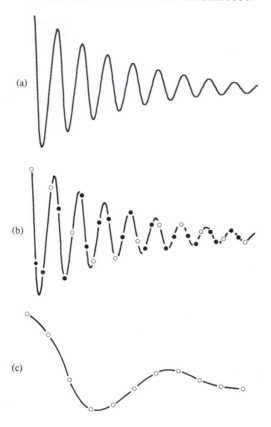

Fig. 7.14 The free induction decay shown in (a) is sampled sufficiently rapidly in (b) to ensure that the decay is interpreted correctly, but is not sampled sufficiently rapidly in (c). The sampling points in (c) are identical to the 1st, 4th, 7th, 10th, etc. points of (b) (open circles); i.e. the sampling frequency in (c) is one-third of the sampling frequency in (b). With this slower sampling frequency the free induction decay shown in (a) is interpreted incorrectly, for its apparent frequency is too low.

detection, the frequency width is twice this value, i.e. $1000/d$ Hz, because both positive and negative frequencies are displayed (see Section 7.5.2). The implications of this are illustrated by the spectra shown in Fig. 7.15. Figure 7.15(a) shows a spectrum obtained using a sampling interval of 330 µs, corresponding to a spectral width of ±1.5 kHz. The signals are all positioned correctly. Figure 7.15(b) shows a spectrum obtained under identical conditions, except that a sampling interval of 500 µs was used, corresponding to a smaller spectral width of ±1 kHz. This is now too long for this particular spectrum, for signal A has a frequency of 1.125 kHz, which is beyond the selected spectral width. As a result, the frequency of signal A is interpreted incorrectly by the sampling process; instead of appearing at the correct

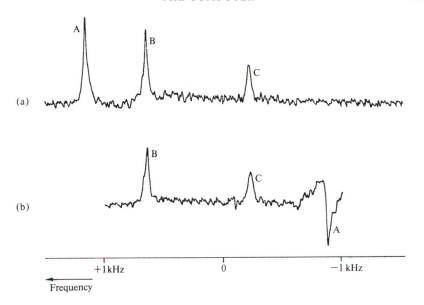

Fig. 7.15 (a) A spectrum obtained using a sampling interval of 330 μs, corresponding to a spectral width of 3 kHz. The signals are all positioned correctly. (b) A spectrum obtained under identical conditions, except that a sampling interval of 500 μs was used. Signal A should be at +1.125 kHz. In (b), the sampling effect causes the signal to be reflected about the end of the spectrum, i.e. about the frequency of 1 kHz, producing what might be expected to be an apparent value of 0.875 kHz. However, there is an additional effect when using quadrature detection that causes the signal to appear, not at +0.875 kHz, but at −0.875 kHz.

frequency of 1.125 kHz, it appears at −0.875 kHz. This effect is known as aliasing and should clearly be avoided, preferably by selecting a sampling interval that corresponds to a sufficiently large spectral width.

Aliasing occurs with noise as well as with signal. For example, if we select a spectral width of ±2.5 kHz, the noise at frequencies beyond ±2.5 kHz will be aliased back into the spectrum to appear as relatively low-frequency noise. It is therefore essential that high-frequency noise be filtered out before entering the computer, otherwise the effect of aliasing would be to increase the noise level significantly and hence to decrease the final signal-to-noise ratio. It is primarily for this reason that the receiver filters are normally set to accept all frequencies within the spectral width but to reject all others. The effects of incorrect filter setting are illustrated in Fig. 7.16. Figure 7.16(a) shows a spectrum obtained under correct conditions; the sweep width is 5 kHz and the filters are set to the same frequency. Figure 7.16(b) shows a spectrum obtained under identical conditions, except that the filters were set to 10 kHz. The resulting aliasing of noise causes the observed increase in the noise level. Figure 7.16(c) shows what happens when the filters are set to 3 kHz; there is

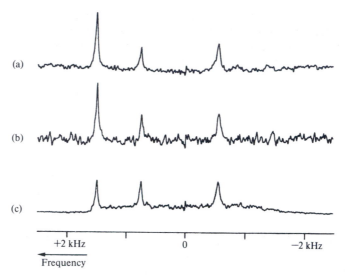

Fig. 7.16 The effects of filter settings: (a) correct setting; (b) filters set to too high a frequency bandwidth result in additional noise; and (c) filters set too too low a bandwidth reduce signal and noise at the extremities of the spectrum.

a large reduction in intensity of both signal and noise at the extremities of the spectrum and these signal intensities no longer reflect the true intensities that should be obtained.

It may be seen from the above discussion that it is the value of the sampling interval that in practice determines the spectral width. Therefore, when an operator specifies the required spectral width, the instrument setting that is actually adjusted is the sampling interval. It may also be seen that, while the filter settings can be set completely independently of the sampling interval, in practice the two settings should match each other; if, for example, a spectral width of ± 2.5 kHz is specified, the filters should correspondingly be set to a cut-off frequency of ± 2.5kHz. Commonly, therefore, on specifying a spectral width in MRS, the filter settings will automatically be adjusted to the appropriate value.

Analogous aliasing effects can also occur in imaging when signal is generated by structures that are beyond the selected field of view. In the frequency-encoding direction, i.e. in the direction that is set by the read gradient, the filters can be set to correspond to the desired field of view, and so aliasing of any structures that are outside the desired field of view can be reduced by filtering out their signals. This is exactly equivalent to the effects in spectroscopy discussed above. However, the filters do not have idealized characteristics; moreover, in the phase-encoding direction there is no analogous filtering procedure. In both directions, therefore, aliasing into the region of interest is best avoided by acquiring data from a sufficiently large field of view.

Another variable that must be set is the number of sampling points in the FID. We recall that Fourier transformation of a given number of data points gives rise to the same number of points in the resulting spectrum or image. As a result, the selection of sampling points in the FID influences (or is influenced by) the required spectral or spatial resolution. Consider, for example, a spectrum of width 5 kHz. If spectral resolution of 1 Hz is required, then adjacent data points in the spectrum must be separated by less than 1 Hz, and so the spectrum must contain at least 5000 data points. The FID must be sampled a correspondingly large number of times. Therefore, for solution studies in particular, when high spectral resolution is required, it is common to acquire data in 4K, 8K (where 1K = 1024), or even more data points per channel. For *in vivo* studies, this large number of sampling points is not normally required, partly because these studies tend to be carried out at lower field strengths and therefore involve correspondingly lower spectral widths, and also because the high degree of spectral resolution that is available in solution studies simply cannot be achieved *in vivo*. One point to note is that the total sampling or acquisition time is the inverse of the frequency separation of points following Fourier transformation. This is another reflection of the inverse relationships between signals in the time and frequency domains, which were discussed in Chapter 5.

As we have seen, imaging studies normally involve a relatively modest number (typically 256) of sampling points in the FID. In addition, because of time constraints associated with the acquisition of many phase-encoding steps, it may be desirable to utilize even fewer points (e.g. 128) in the phase-encoding direction. However, if a 256×256 pixel image is required, this can nevertheless be achieved with the acquisition of a 128×256 data set (i.e. with 128 phase-encoding steps), by taking advantage of a procedure known as zero filling (see Section 7.6.2.1).

7.6.2 Data processing

Following accumulation of the FID, the resulting signal must be processed in order to produce a recognizable spectrum or image. In addition to the process of Fourier transformation, a number of additional steps are usually required, and in this section we discuss briefly some important aspects of data processing. Some more sophisticated automated methods are described in the following section.

7.6.2.1 *Zero filling*

A procedure known as zero filling is sometimes carried out prior to Fourier transformation. This process involves appending an array of zeros to the end of the acquired array of data. This has the effect of increasing the number of points in the transformed data set; i.e. it provides a means of interpolation, resulting in a spectrum or image that may have a smoother appearance.

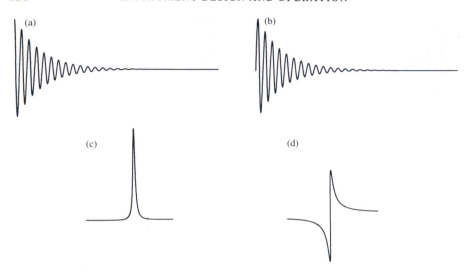

Fig. 7.17 Free induction decays observed with quadrature detection in the two channels (a) and (b), together with the signals obtained on Fourier transformation. The two signals (c) and (d) represent the real and imaginary components. (c) is the form in which the spectra are finally displayed.

7.6.2.2 *Phase correction*

In Section 7.5.2, it was pointed out that the NMR receiver incorporates quadrature detection, as a result of which two free induction decays, 90° out of phase with each other, are observed in two separate channels of the receiver system. Figure 7.17 shows free induction decays that may be observed in the two channels in the case of a simple spectroscopy study (note that commonly only one is actually displayed on the computer screen), together with the results of Fourier transformation. The spectrum obtained on Fourier transformation also contains two components, known as the real and imaginary terms, which in Fig. 7.17 correspond to the absorption and dispersion modes (see Appendix to Chapter 5, p. 164). The absorption mode is the form that characterizes the absorption of energy by the nuclear spins, and in this idealized case has a Lorentzian lineshape. It is in this form that spectra are finally displayed. In practice, however, the phase of the FID and of its resulting Fourier transform is determined by the precise way in which the NMR instrument is set up, and is somewhat arbitrary; hence the FID and its Fourier transform might more generally be of the form shown in Fig. 7.18. Conversion to the required form of Fig. 7.17 is accomplished by applying the appropriate phase correction to the transformed spectrum.

There are several factors that result in the need for phase correction. Consider, for example, the effect of a finite delay time between the end of the radiofrequency pulse and the start of data acquisition. Such a delay is

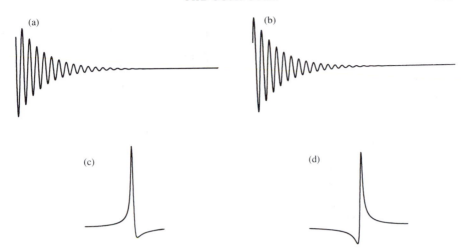

Fig. 7.18 Free induction decays observed with quadrature detection in the two channels (a) and (b), but with an arbitrary phase, as illustrated by the fact that the initial points in the decays are neither zero nor at a maximum or minimum. Fourier transformation gives rise to real and imaginary components as shown in (c) and (d).

often required to ensure that the radiofrequency coil and receiver have recovered from the direct effects of the high-voltage radiofrequency pulses. If this delay time (which is typically similar to the sampling interval) is t, then in this time a waveform of frequency v goes through vt cycles and suffers a phase shift of $2\pi vt$ radians. A phase correction of this amount is therefore introduced as a result of this delay. It should be noted that resonances of different frequencies undergo different phase shifts; in fact the delay introduces a phase shift that is directly proportional to frequency. This can be corrected for by the so-called frequency-dependent phase correction. The use of a constant phase correction, together with a phase correction that is linearly proportional to frequency, should ensure that all resonances within a spectrum are correctly phased.

The phase correction makes use of both real and imaginary components of a spectrum. A phase change of ϕ corresponds to manipulation of each data point according to the following equations:

$$r_2 = - i_1 \sin \phi + r_1 \cos \phi$$

$$i_2 = r_1 \sin \phi + i_1 \cos \phi$$

where r_2 and i_2 correspond to the real and imaginary components of each point of the spectrum following the phase correction and r_1 and i_1 correspond to the points prior to correction. It is therefore essential to preserve both real and imaginary components of the spectrum until phase correction is complete.

The production of images often bypasses the need for phase correction, by displaying the signal in the magnitude (or absolute value) mode, as the modulus M of the real and imaginary components:

$$M = \sqrt{r^2 + i^2}$$

where r and i represent the amplitudes of the real and imaginary components. One of the problems with presenting data in this form is that, on Fourier transformation of a free induction decay, it is found that the 'wings' of the signals are much more extensive following magnitude correction than following phase correction and display in the absorption mode. This can have severely adverse effects on spectral resolution, and magnitude corrections are therefore avoided in spectroscopy, if at all possible. However, if Fourier transformation is carried out on a signal that includes both the growing and the decaying part of an echo, it may be shown that, provided these two parts of the echo fall off symmetrically about the centre, the magnitude correction produces a pure absorption signal, eliminating the problems generated by the 'wings'. This is one of the advantages of acquiring a full echo, rather than just the decaying half. Of course, if necessary (for example when constructing phase images in blood flow studies) phase corrections can be carried out in imaging studies, using techniques that are similar to those used for spectroscopy.

7.6.2.3 *Manipulation of the FID*

The quality of NMR data is ultimately determined by the signal and noise that are stored within the computer. However, certain characteristics of the resulting images and spectra can be considerably improved by means of appropriate manipulation of the data prior to Fourier transformation. In this section, we illustrate the basic principles underlying some of the methods of data manipulation; again, for simplicity we use examples from spectroscopy.

Sensitivity enhancement   Consider a simple FID and its transform as shown in Fig. 7.19(a). Suppose we wish to improve the signal-to-noise ratio of the spectrum. This can be achieved by multiplying the FID point by point by a decaying exponential as shown in Fig. 7.19(b). The function of this multiplication is to eliminate the noise preferentially by lending more weight to the initial part of the FID, where the signal-to-noise ratio is high, than to the latter part, where the signal-to-noise ratio is much lower. The effects of exponential multiplication are shown in Fig. 7.19(c). There are two important features to note when comparing the transformed spectrum with that shown in Fig. 7.19(a). First, the signal-to-noise ratio is considerably enhanced by performing the multiplication. Secondly, however, the linewidth is increased as a result of this process. The reason for this is that multiplication by the decaying exponential has the effect of reducing the time constant of the FID. As we have seen, there is a reciprocal relationship between the time constant of the decay and the linewidth, and therefore the linewidth increases correspondingly. In fact, the exponential multiplication increases the linewidth by an amount $\Delta v = 1/\pi T$, where T is the time constant of the exponential decay.

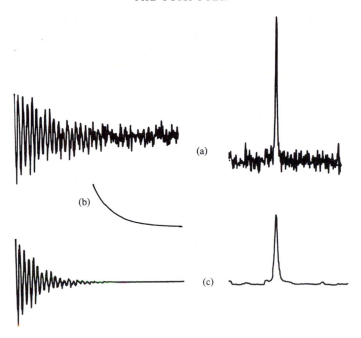

Fig. 7.19 (a) A free induction decay and its Fourier transform. The free induction decay in (c) is obtained on multiplying the decay of (a) point by point by the exponential decay of (b). This results in a signal of enhanced signal-to-noise, but also of enhanced linewidth.

It can be seen from Fig. 7.19 that exponential multiplication in the time domain has the same effect as filtering would have in the frequency domain, and in fact the process is sometimes referred to as filtering. One of the virtues of Fourier-transform NMR is the ease with which this process is achieved, for filtering in the frequency domain is rather more complicated. (It should be noted that this filtering process is completely different from the function of the filters used in the spectrometer receiver.)

For an optimal signal-to-noise ratio in the transformed spectrum, the decaying exponential should have the same time constant as the FID. The effect of this so-called 'matched filter' is to double the linewidth, and therefore the resolution is significantly degraded by the process. If a spectrum contains several resonances of differing linewidths, it is clearly impossible to optimize the time constant of the exponential multiplication for all of the lines. Under these circumstances the spectrum could be processed several times using different time constants in each case. Usually, however, this should be unnecessary as the signal-to-noise ratio achieved in the final spectrum is not very sensitive to mis-setting of the time constant of the multiplication.

Analogous filtering techniques may also be applied to imaging data, so that, for example, signal-to-noise ratios can be enhanced but at the expense of some spatial resolution.

Resolution enhancement Figure 7.19 illustrates the important rule that enhancement of the signal-to-noise ratio can only be achieved at the expense of resolution. The reverse of this statement also holds; i.e. resolution enhancement is only achieved at the expense of signal-to-noise. Essentially, resolution-enhancement methods work by applying mathematical manipulations to the FID that selectively enhance those components of the FID (the long-lived components) that tend to give narrow signals. A wide range of weighting functions have been proposed.

The simplest method is to multiply the FID by a growing exponential rather than by the decaying exponential of Fig. 7.19(b). However, this method can lead to problems; in particular, there will be distortion of the Fourier-transformed spectrum if the signal is non-zero at the end of the period of data acquisition. For this and other reasons, a variety of alternative techniques for resolution enhancement have been devised, each of which imposes a different shaping function upon the FID (see Lindon and Ferrige 1980; Freeman 1988). For example, the Lorentzian-to-Gaussian conversion uses a shaping function of the form $\exp[at/T_2^* - b(t/T_2^*)^2]$, which converts a Lorentzian lineshape to a Gaussian lineshape if the parameter a is chosen appropriately. As shown in Fig. 7.20, in comparison with Lorentzian lines, Gaussian lines have the advantage that a far smaller percentage of signal is in the 'wings'. Another method is termed convolution difference (Campbell *et al.* 1973). In its simplest form, it involves multiplying the FID by a decaying exponential, subtracting the processed FID from the original unprocessed FID, and then finally reprocessing the resulting signal in the standard manner. Effectively, this procedure subtracts out the broad spectral components, while leaving the narrow components *relatively* unaffected. However, the remaining narrow components are necessarily distorted by the procedure; in particular, they will have negative-going wings (Fig. 7.21). In fact, it is interesting to note that the total integrated intensity of any signal that is obtained following Fourier transformation is proportional to the *initial* amplitude of its corresponding FID; following the convolution difference procedure described above, the initial amplitude of the processed FID is zero, and in practice this means that the intensities of the negative-going wings will be equal and opposite to the intensity of the positive central component of the signal.

Comparison of the ^{31}P spectra of Fig. 2.9(b) and (c) shows how resolution-enhancement techniques can be used in practice to sharpen up metabolite signals and remove underlying broad spectral components.

7.6.3 Automated methods of data analysis

Several computer methods are now available for automated processing and analysis of NMR data (Stephenson 1988; de Beer and van Ormondt 1992). Essentially, these methods fit the measured data to a series of spectral lines, and provide a means of determining signal intensities that avoids the need for the more operator-dependent 'manual' methods. The computer analysis can

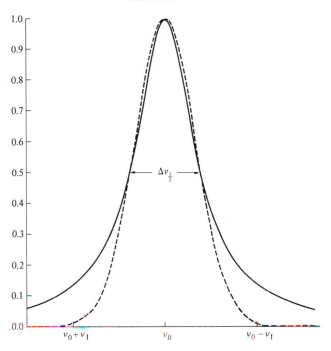

Fig. 7.20 A Lorentzian NMR signal (solid line), together with a Gaussian (dashed line) of the same width and height.

be carried out in the frequency domain, i.e. following Fourier transformation of the free induction decay, or alternatively in the time domain, i.e. by direct fitting of the free induction decay itself. This is an area of active research, but the automated methods necessarily have problems when dealing with data that are of intrinsically poor quality; ultimately there is no substitute for obtaining raw data that, given the various constraints of each individual NMR study, are of the highest possible quality.

REFERENCES

Ackerman, J. J. H., Grove, T. H., Wong, G. G., Gadian, D. G., and Radda, G. K. (1980). Mapping of metabolites in whole animals by ^{31}P NMR using surface coils. *Nature*, **283**, 167–70.

Becker, E. D., Ferretti, J. A., and Gambhir, P. N. (1979). Selection of optimum parameters for pulse Fourier transform nuclear magnetic resonance. *Anal. Chem.*, **51**, 1413–20.

de Beer, R. and van Ormondt, D. (1992). Analysis of NMR data using time domain fitting procedures. In *NMR basic principles and progress*, (eds P. Diehl, E. Fluck, H. Günther, R. Kerfeld, and J. Seelig) vol. 26, pp. 201–48.

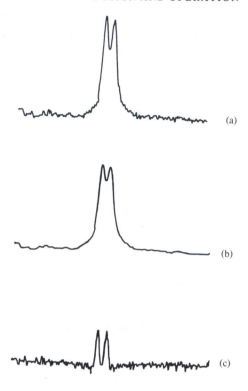

Fig. 7.21 Resolution enhancement by the convolution difference technique (Campbell *et al.* 1973). (a) The Fourier transformation of an FID. (b) The Fourier transformation of an FID that had been multiplied by a decaying exponential. Subtraction of (b) from (a) gives (c). (c) clearly has better spectral resolution than (a), but poorer signal-to-noise. Note also that there are very broad negative-going wings to the lines in (c), which result from the convolution difference process.

Bosch, S. and Ackerman, J. J. H. (1992). Surface coil spectroscopy. In *NMR basic principles and progress*, (eds P. Diehl, E. Fluck, H. Günther, R. Kerfeld, and J. Seelig) vol. 27, pp. 3–44.

Campbell, I. D., Dobson, C. M., Williams, R. J. P., and Xavier, A. V. (1973). Resolution enhancement of protein PMR spectra using the difference between a broadened and a normal spectrum. *J. Magn. Reson.*, **11**, 172–81.

Chen, C. N., Hoult, D. I., and Sank, V. J. (1983) Quadrature detection coils: a further $\sqrt{2}$ improvement in sensitivity. *J. Magn. Reson.*, **54**, 324–7.

Ernst, R. R. and Anderson, W. A. (1966). Application of Fourier transform spectroscopy to magnetic resonance. *Rev. Sci. Instrum.*, **37**, 93–102.

Freeman, R. (1988). *A handbook of magnetic resonance*. Longman Scientific and Technical, Harlow.

Gadian, D. G. and Robinson, F. N. H. (1979). Radio-frequency losses in NMR experiments on conducting samples. *J. Magn. Reson.*, **34**, 449–55.

Garwood, M. and Ugurbil, K. (1992). B_1 insensitive adiabatic RF pulses. In *NMR basic principles and progress*, (eds P. Diehl, E. Fluck, H. Günther, R. Kerfeld, and J. Seelig) vol. 26, pp. 109–47.

Hayes, C. E., Edelstein, W. A., Schenck, J. F., Mueller, O. M., and Eash, M. (1985). An efficient highly homogeneous radiofrequency coil for whole body NMR imaging at 1.5 T. *J. Magn. Reson.*, **63**, 622–8.

Hoult, D. I. (1978). The NMR receiver: a description and analysis of design. *Prog. NMR Spectrosc.*, **12**, 41–77.

Hoult, D. I. and Lauterbur, P. C. (1979). The sensitivity of the zeugmatographic experiment involving human samples. *J. Magn. Reson.*, **34**, 425–33.

Hoult, D. I. and Richards, R. E. (1975). Critical factors in the design of sensitive high resolution nuclear magnetic resonance spectrometers. *Proc. R. Soc. Lond. A*, **344**, 311–40.

Hoult, D. I. and Richards, R. E. (1976). The signal-to-noise ratio of the nuclear magnetic resonance experiment. *J. Magn. Reson.*, **24**, 71–85.

Kanal, E., Shellock, F. G., and Talagala, L. (1990). Safety considerations in MR imaging. *Radiology*, **176**, 593–606.

Lindon, J. C. and Ferrige, A. G. (1980). Digitisation and processing in Fourier transform NMR. *Prog. NMR Spectrosc.*, **14**, 27–66.

Link, J. (1992). The design of resonator probes with homogeneous radiofrequency fields. In *NMR basic principles and progress*, (eds P. Diehl, E. Fluck, H. Günther, R. Kerfeld, and J. Seelig) vol. 26, pp. 1–31.

Meakin, P. and Jesson, J. P. (1973). Computer simulation of multipulse and Fourier transform NMR experiments. I. Simulations using the Bloch equations. *J. Magn. Reson.*, **10**, 290–315.

Morris, P. G. (1992). Frequency-selective excitation using phase-compensated RF pulses in one and two dimensions. In *NMR basic principles and progress*, (eds P. Diehl, E. Fluck, H. Günther, R. Kerfeld, and J. Seelig) vol. 26, pp. 149–70.

Morse, O. C. and Singer, J. R. (1970). Blood flow measurements in intact subjects. *Science*, **170**, 440–1.

Redfield, A. G. and Kunz, S. (1975). Quadrature Fourier NMR detection: simple multiplex for dual detection and discussion. *J. Magn. Reson.*, **19**, 250–4.

Schnall, M. (1992). Probes tuned to multiple frequencies for in-vivo NMR. In *NMR basic principles and progress*, (eds P. Diehl, E. Fluck, H. Günther, R. Kerfeld, and J. Seelig) vol. 26, pp. 33–63.

Shaw, D. (1976). *Fourier transform n.m.r. spectroscopy*. Elsevier-North-Holland, Amsterdam.

Shellock, F. G. and Kanal, E. (1994). Guidelines and recommendations for MR imaging safety and patient management: III. Questionnaire for screening patients before MR procedures. *J. Magn. Reson. Imag.*, **4**, 749–51.

Stejskal, E. O. and Schaeffer, J. (1974). Comparisons of quadrature and single-phase Fourier transform NMR. *J. Magn. Reson.*, **14**, 160–9.

Stephenson, D. S. (1988). Linear prediction and maximum entropy methods in NMR spectroscopy. *Prog. NMR Spectrosc.*, **20**, 515–626.

Turner, R. (1993). Gradient coil design: a review of methods. *Magn. Reson. Imag.*, **11**, 903–20.

van Zijl, P. C. M. and Moonen, C. T. W. (1992). Solvent suppression strategies for *in vivo* magnetic resonance spectroscopy. In *NMR basic principles and progress*, (eds P. Diehl, E. Fluck, H. Günther, R. Kerfeld, and J. Seelig) vol. 26, pp. 67–108.

8

Pulse sequences

The development of new pulse sequences has been a major theme of research in magnetic resonance for many years, and has provided much of the basis for the continued progress in imaging and spectroscopy. For example, the studies described in Chapter 4 illustrate the various ways in which specialized imaging sequences can be used, not only for the visualization of specific structures, but also for the investigation of tissue function and pathophysiology. In this final chapter, we discuss in more detail some of the more widely used pulse sequences. We show that these sequences, which involve appropriate combinations of radiofrequency pulses and pulsed magnetic field gradients, can generally be built up on a modular basis, and that pulse sequences used for spectroscopy have many features in common with those used for imaging.

8.1 SIMPLE DATA ACQUISITION

The simplest pulse sequence can be referred to as 'pulse and acquire', and is illustrated in Fig. 8.1. A radiofrequency pulse is applied, and the response (i.e. free induction decay) is collected. This process is normally repeated in order to increase the signal-to-noise ratio; the summation of N free induction decays leads to a signal-to-noise improvement of $\sqrt{N}$, once a steady state has been reached. On Fourier transformation, a spectrum is obtained, the characteristics of which depend upon a number of factors, including the chemical composition of the sample and the static field homogeneity (the more homogeneous the field, the narrower the lines).

The one-dimensional imaging analogue of this sequence is shown in Fig. 8.2. In this case, the application of the magnetic field gradient G_x causes the frequency of the signal to become spatially dependent, generating a response that gives information about the distribution of the sample along the x-axis. If the gradient is sufficiently large, it will overwhelm chemical shift effects, and all chemical information will be lost. If the frequency shifts associated with the chemical shifts are comparable with those induced by the gradients, then there will be a complex superimposition of the two effects which is difficult to disentangle, and which may result in so-called 'chemical shift artefacts' in imaging studies.

The apparent displacement of lipid-rich structures provides a good illustration of chemical shift artefacts. In conventional MRI studies, any frequency

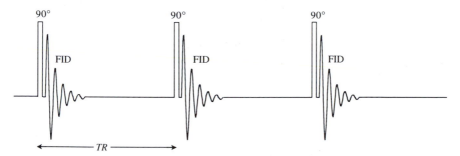

Fig. 8.1 A simple 'pulse and acquire' sequence in which 90° pulses are applied at intervals of *TR*. For simplicity, the signals that are shown in this and the following figures represent just a single frquency component; in practice the signals would normally be a lot more complex.

shifts between signals are automatically attributed to differences in spatial location rather than to chemical shift effects. Therefore the chemical shift difference of about 3 ppm (i.e. 200 Hz at 1.5 T) between the water and lipid signals manifests itself as a spatial displacement of the lipid protons relative to the water protons. A simple calculation shows the magnitude of this effect. In an imaging study, a shift of 200 Hz would be generated by a displacement of 1 mm if data were acquired in the presence of a read gradient of 4.6 mT m^{-1} (as this field gradient gives rise to a ^{1}H frequency gradient of 200 Hz mm^{-1}). Therefore, if imaging data are acquired at 1.5 T using a read gradient G_x of 4.6 mT m^{-1}, then the chemical shift difference between the lipid and water protons will result in an apparent displacement of 1 mm

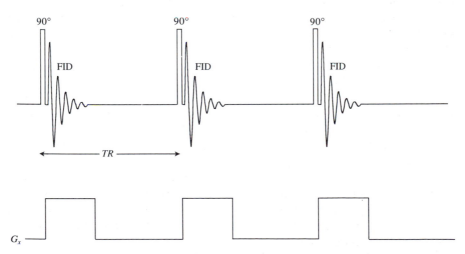

Fig. 8.2 A one-dimensional imaging sequence in which signal is acquired in the presence of a read gradient G_x (but see Section 8.8 for the practical use of the read gradient in MRI).

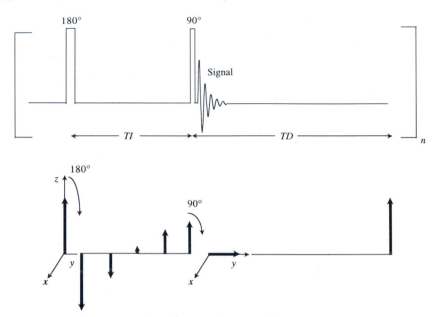

Fig. 8.3 The inversion recovery sequence for the measurement of T_1. The displayed sequence may be repeated n times at each value of TI to build up the signal-to-noise ratio. The heavy arrows indicate how the magnetization changes during the sequence.

of the lipid protons relative to the water protons. To avoid any misinterpretation that may arise, in some circumstances it may be necessary to obtain images in which the fat signal is suppressed, for example by selective saturation (see Section 8.5).

8.2 T_1-WEIGHTING AND THE MEASUREMENT OF T_1

If the radiofrequency pulses in the sequence of Fig. 8.1 are repeated at intervals TR that are similar to, or shorter than, the T_1 values of the nuclei of interest, then the response to each pulse is reduced in intensity through the process of saturation (see Section 7.3.2). This provides a simple basis for T_1-weighted contrast, the signal intensity being dependent upon the pulse angle and on the ratio TR/T_1.

More sophisticated sequences can be used for the generation of T_1-weighted contrast, and for the measurement of T_1 values. The most well known of these sequences is the inversion recovery sequence shown in Fig. 8.3. The 180° pulse causes inversion of the magnetization, which then gradually recovers towards its equilibrium value M_0 with the time constant T_1. After a given time TI, a 90° pulse is applied. This effectively provides a 'snapshot' of the

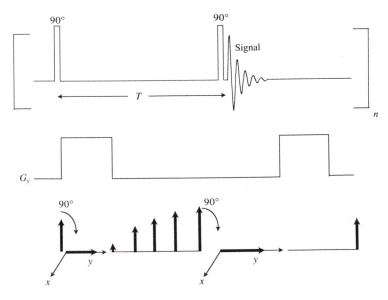

Fig. 8.4 The saturation recovery sequence for the measurement of T_1. The heavy arrows indicate how the magnetization changes during the sequence.

value of the magnetization at time TI, and by repeating the sequence with different TI values, a range of snapshots can be obtained. The value of T_1 can then be obtained from a plot of $\ln(M_\infty - M_{TI})$ against TI, which should be a straight line of gradient $1/T_1$. If protons in different tissues have different T_1 values, then they will recover at different rates. At a given TI value, they will therefore generate signals of different intensities, thereby generating T_1-weighted contrast. This $180°-TI-90°$ sequence can be regarded as a module which can be incorporated into more complex imaging and spectroscopy sequences in order to produce appropriate T_1-weighting.

A feature of the inversion recovery method is that the measurement of T_1 is little affected by any inaccuracies in the $180°$ and $90°$ pulses that may result from incorrect setting of the pulse angles or, more commonly, from inhomogeneities in the B_1 field. Another advantage of the method is that it has twice the dynamic range of other techniques, in that signals can vary in intensity from $-M_0$ to $+M_0$, rather than from 0 to $+M_0$. However, it is a time-consuming method, because each $90°$ pulse is followed by a long delay (typically about $4T_1$) in order to allow the system to return to equilibrium between consecutive acquisitions. This is a major drawback when examination time is limited. An alternative method, known as saturation recovery, is shown in Fig. 8.4. In this method a $90°$ pulse tilts the magnetization into the xy-plane and then a spoiler pulse is applied. This is a pulsed field gradient that destroys the B_0 homogeneity for a brief period and hence destroys the

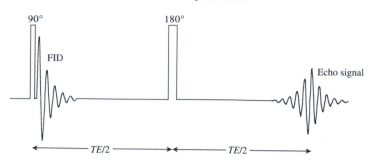

Fig. 8.5 The Hahn spin-echo sequence.

net magnetization in the *xy*-plane. The overall effect should be that there is no resulting magnetization along any direction. After a time T a second 90° pulse is applied and the signal is monitored, another spoiler pulse is applied, for the same reason as previously, and the whole process is repeated for a range of values of T. Again, a suitable semi-logarithmic plot enables T_1 to be measured. The main advantage of the method is that T_1 values can usually be measured much more rapidly than with the inversion recovery method; this is because the long delay time between consecutive acquisitions is not required.

If speed of measurement is to be the major criterion for choice of method, then the progressive saturation technique should be considered. Essentially, this just involves applying a train of 90° pulses and collecting signal in the conventional manner. After a sufficient signal-to-noise ratio has been acquired at a given time interval, the interval is changed and the process is repeated. Again, a semi-logarithmic plot gives T_1. The main problem with the method is that the measured T_1 value is sensitive to the accuracy of the 90° pulses; however, increasingly sophisticated methods are becoming available for generating accurate pulses. Also, it is essential to ensure that the magnetization reaches a steady state prior to the start of each accumulation, and for this reason signal should not be acquired from the first of a train of pulses.

8.3 T_2-WEIGHTING AND THE MEASUREMENT OF T_2

The spin-echo sequence (Hahn 1950) is extensively used both in imaging and spectroscopy, and generates signals that are weighted according to T_2. In its simplest form, it consists of a 90° pulse followed, after a delay time of $TE/2$, by a 180° pulse (Figs 8.5 and 8.6). The 90° pulse tilts the magnetization into the *xy*-plane, or alternatively into the $x'y'$-plane of the rotating frame of reference; if the pulse is applied about the x' axis, then the magnetization aligns along the y' axis (Fig. 8.6(b)). If all the spins were to precess at exactly

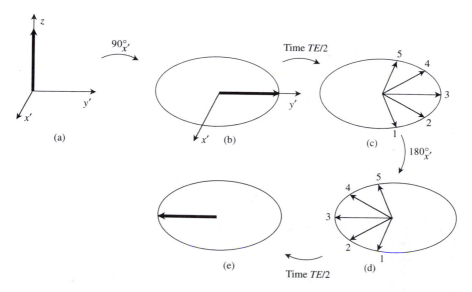

Fig. 8.6 Diagram showing the fanning out and refocusing of magnetization in the course of the spin-echo sequence of Fig. 8.5. In this case both the 90° and 180° pulses are applied about the x'-axis of the rotating frame of reference, and the magnetization refocuses along the $-y'$-axis.

the same frequency, then in the rotating frame of reference the $x'y'$-magnetization generated by the 90° pulse would remain static along the y'-axis. However, as a result of B_0 inhomogeneity, nuclei in different regions of the sample experience slightly different fields and therefore precess at slightly different frequencies. In the rotating frame of reference this will result in a 'fanning out' of the magnetization as shown in Fig. 8.6(c) (see also frontispiece). There is a gradual loss of coherence, and therefore of signal, as observed in a free induction decay. Now if a 180° pulse is applied about the x'-axis at a time $TE/2$, the various components of the magnetization will be inverted as shown, and provided that each component continues to precess at the same frequency (as it will if it stays in the same field strength) all the components of magnetization will arrive at the $-y'$-axis at the same time (Fig. 8.6(e)). This results in the formation of an 'echo' signal at a time $TE/2$ after the 180° pulse. Since the echo is formed along the negative y'-axis, the signal will be of opposite phase to that observed after the 90° pulse. The echo should have the same amplitude as the initial signal if no significant relaxation occurs during the time TE and if the effects of diffusion of molecules during this time are small (see Section 8.5). However, if there is any dephasing or 'fanning out' of the magnetization during the time TE as a result of T_2 relaxation processes, this cannot be refocused by the 180° pulse because of the random nature of these processes. Therefore the amplitude of the echo

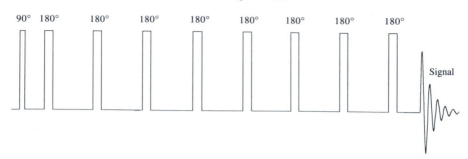

Fig. 8.7 The Carr–Purcell–Meiboom–Gill sequence.

will be diminished by a factor e^{-TE/T_2}. Thus T_2 can be evaluated from measurements of the echo intensities obtained at a range of different *TE* values.

As discussed in earlier chapters, signal loss during the period *TE* can also occur through additional mechanisms, for example through the effects of diffusion or bulk flow. In practice, diffusion- or flow-dependent effects can be specifically emphasized in order to produce diffusion or flow weighting (see Section 8.5). Alternatively, these effects can be reduced through modifications to the pulse sequences. For example, the effective diffusion time can be reduced by applying a train of 180° pulses during the time *TE*, rather than just one 180° pulse. Figure 8.7 shows such a pulse train. Each 180° pulse generates an echo as described above, and so if signals were detected midway between each pulse a train of echoes would be observed (Carr and Purcell 1954). In Fig. 8.7, just the final echo is displayed. In a simple modification to this pulse train, the 180° pulses are applied about the y'-axis of the rotating frame rather than about the x'-axis (corresponding to a phase shift of 90°). This has two important effects: first the echoes all form along the same direction rather than alternately along the y'- and $-y'$-axes and, secondly, any errors in the accuracy of the 180° pulses tend to cancel out rather than to add as they would do in the absence of the phase shift. This modification, which was proposed by Meiboom and Gill (1958), results in what is now known as the Carr–Purcell–Meiboom–Gill (CPMG) sequence. It may be shown that the echo train undergoes a gradual loss of signal intensity through T_2 effects. However, the signal losses due to diffusion are markedly reduced, because of the reduced diffusion time between the consecutive 180° pulses. As a result, the Carr–Purcell–Meiboom–Gill sequence tends to give more accurate T_2 measurements than the simple $90°–TE/2–180°–TE/2$ spin-echo sequence.

The 90° phase shift that was proposed by Meiboom and Gill is equally applicable to the simple Hahn spin-echo sequence; if the 180° pulse is applied about the y'-axis rather than the x'-axis, then it may readily be seen that

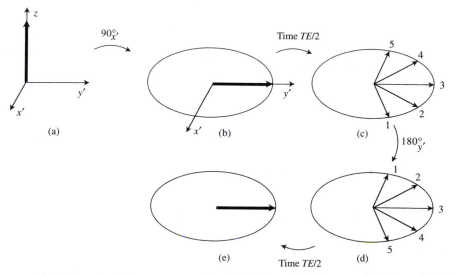

Fig. 8.8 Hahn spin-echo sequence with a 90° phase shift between the 90° and 180° pulses. In contrast to Fig. 8.6, the magnetization now refocuses along the y'-axis.

the echo forms along the $+y'$-axis rather than about the $-y'$-axis (Fig. 8.8). More generally, phase shifts in the radiofrequency pulses provide a powerful means of influencing both wanted and unwanted signals, and feature prominently in commonly used pulse sequences. A simple example was given in Section 7.5.3, where it was pointed out that alternation of the phase of the radiofrequency excitation by 180° in consecutive acquisitions can eliminate artefactual signals that might otherwise appear at zero frequency.

Spin-echo sequences are extensively used in imaging and spectroscopy, both for routine acquisition of data and also for more specialized applications, as discussed in some of the following sections.

8.4 T_2^*-WEIGHTING AND THE MEASUREMENT OF T_2^*

Just as T_2 can be measured from a series of spin-echo measurements, so T_2^* can be measured by means of gradient echoes. Gradient echoes are now extensively used, for example in rapid imaging sequences and in functional imaging studies, and in this section we briefly discuss this type of echo.

In general, echoes are generated by processes that reverse the dephasing of magnetization. In the spin-echo sequence, the 180° pulse reverses the effects of local field inhomogeneities and hence generates a T_2-weighted spin-echo signal. An alternative means of generating an echo is by reversal of an applied magnetic field gradient, as shown in Fig. 8.9. The first (negative) gradient gradually dephases the magnetization, by an amount that is dependent upon

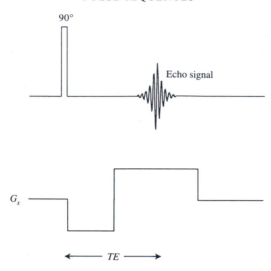

Fig. 8.9 A simple gradient echo generated by the reversal of an applied field gradient G_x.

the integral of the gradient over time. On reversal of the gradient, the magnetization gradually rephases, the maximum rephasing occurring when this integral is fully reversed. At this time, the signal is regenerated, producing a so-called gradient echo. The main difference between a spin echo and a gradient echo is that the gradient echo does not refocus the dephasing effects of intrinsic field inhomogeneities, and therefore generates a signal that is weighted according to T_2^* rather than T_2. Signals in gradient echo images may therefore be attenuated in regions where the field homogeneity is poor, such as at interfaces between tissue and air. While this is a disadvantage, a similar effect is put to good use in functional neuroimaging, which specifically exploits the differential local field gradients generated by deoxyhaemglobin in the blood. More generally, gradient echoes are commonly used for rapid imaging, as the echo time TE can be made much shorter than in spin-echo studies. Under such circumstances (i.e. at short TE values), field inhomogeneity effects produce relatively little signal loss.

8.5 DIFFUSION WEIGHTING

It was mentioned above that conventional Hahn spin-echoes may be reduced in intensity as a result of the effects of molecular diffusion during the time period TE. While this reduction in intensity is often an undesirable effect, in some situations it may be of interest to use such effects to investigate the diffusional properties of water or other molecules. Indeed, as discussed in

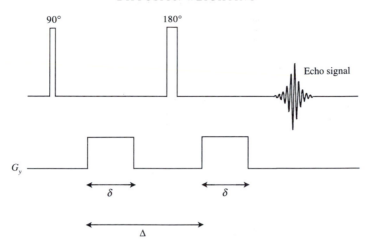

Fig. 8.10 A spin-echo sequence with the incorporation of gradients G_y that generate diffusion weighting.

Section 4.5, diffusion-weighted imaging is now providing important new information about brain pathophysiology. The effects of diffusion can be specifically highlighted by incorporating strong pulsed gradients into the standard spin-echo sequence, as shown in Fig. 8.10. The basic mechanism underlying this signal attenuation is straightforward. The full echo intensity is only achieved if the nuclei experience the same fields following the 180° pulse as they do before it. If the nuclei were not to move at all during the period TE, then the dephasing effects of the second pulsed gradient would cancel out those of the first, and the nuclei would end up in phase at the time of the echo, with no attenuation of the signal other than that due to intrinsic relaxation effects. However, if the nuclei undergo random translational motion as a result of molecular diffusion, then each nucleus will tend to experience different field strengths during the second pulsed gradient from those experienced during the first, and as a result there will be incomplete rephasing. The signal attenuation due to diffusion is related to the gradient strength and duration, and to the diffusion time. In fact, it may be shown that the echo signal is attenuated by a factor

$$A = \exp(-bD) \qquad (8.1)$$

where D is the diffusion coefficient. The factor b is dependent on the gradient strength and timing, and is given by

$$b = \gamma^2 \delta^2 G^2 (\Delta - \delta/3) \qquad (8.2)$$

where γ is the gyromagnetic ratio, δ and G are the duration and amplitude of the diffusion-sensitizing gradient pulses, and Δ is the time interval between the leading edges of these pulses.

Diffusion-weighted imaging can be carried out simply by incorporating the pulsed field gradients into conventional imaging sequences. However, because the image intensities are sensitive to the very small movements that are characteristic of diffusion, they are equally sensitive to any other mechanisms whereby small movements may occur, and as a result there can be considerable problems with motion artefacts. Therefore, considerable effort is being put into the development of methods for minimizing such artefacts.

8.6 THE MEASUREMENT OF FLOW

As discussed in earlier chapters, the dependence of magnetic resonance signals on flow effects makes it possible to obtain images that selectively display the vasculature. While flow can influence magnetic resonance signals in numerous ways (see Bradley 1992; Masaryk *et al.* 1992), we concentrate here on the two main ways in which flow effects are exploited.

As we have seen (Section 4.3), time-of-flight angiography relies on the fact that, if consecutive excitation pulses are applied at time intervals much shorter than T_1, blood flowing into a selected region of interest will be saturated to a lesser extent than stationary tissue. This inflow effect simply reflects the fact that stationary spins experience all of the excitation pulses, while spins in blood flowing into the region of interest do not. The vasculature therefore shows up brightly against the adjacent tissue, and can be displayed selectively with the aid of the post-processing routine known as maximum intensity projection. The pulsing parameters are chosen with a view to optimizing the differential saturation of the stationary and flowing protons, but they are also influenced by the second important effect of flow, namely the effect of phase changes that are associated with pulsed field gradients. While these phase changes should generally be minimized for time-of-flight studies, they are specifically exploited in phase-sensitive angiographic techniques. An understanding of the effects of phase is therefore of relevance to both approaches to magnetic resonance angiography.

We saw in Chapter 5 that the application of a pulsed field gradient G_y causes spins to undergo a phase change ϕ given by

$$\phi = \gamma G_y yt \qquad (8.3)$$

The phase shift is proportional to the y-coordinate of the nuclei, the gradient strength, and the time for which the gradient is applied. If a bipolar field gradient is applied (see Fig. 8.11), in which a negative gradient is followed by a positive gradient of equal magnitude and duration, then stationary nuclei undergo no net phase shift, as the effects of the positive and negative gradients cancel out. However, this cancellation does not occur for spins flowing along the direction of the gradient, since the field strengths experienced during the negative gradient are not equal and opposite to those experienced during the positive gradient.

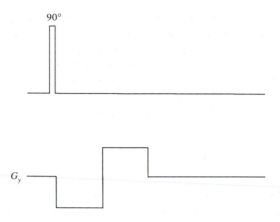

Fig. 8.11 A bipolar field gradient G_y applied after a 90° pulse. Following such a gradient, there is no net phase shift for stationary spins, but moving spins undergo a phase shift.

Let us consider this in more detail. If nuclei move along the direction of a gradient G_y with velocity v_y, their y-coordinate can be written $y = y_0 + v_y t$, where y_0 is the y-coordinate at time zero. Equation (8.3) can then be modified to become

$$\phi = \int_{t_1}^{t_2} \gamma G_y (y_0 + v_y t) \, dt \tag{8.4}$$

and the resulting phase change is given by

$$\phi = \gamma G_y y_0 (t_2 - t_1) + \gamma G_y v_y \frac{(t_2 - t_1)^2}{2} \tag{8.5}$$

The first of these terms corresponds to the expression given in eqn (8.3), and as such cancels out when a positive gradient is followed by an equal and opposite negative gradient. However, it may readily be shown that with this type of bipolar gradient pulse, the second term does not cancel out; instead, it gives rise to a residual phase shift that is proportional to the velocity and also depends upon the gradient strengths and timings.

In practice, these phase shifts have two main effects on images. First, spins flowing at a given velocity all undergo the same phase shift which, as discussed below, is exploited in phase-contrast angiography. Secondly, the phase shifts can lead to signal loss. For example, in laminar flow there is a velocity gradient across a vessel; spins at the centre of the vessel flow faster than those near the walls, and as a result the spins within a single pixel may undergo a range of different phase shifts. The resulting 'fanning out' of the nuclear spins within each pixel leads to signal loss. This loss of signal can create problems, but it

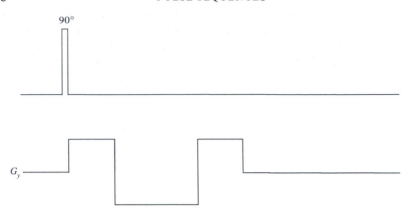

Fig. 8.12 Additional gradient pulses can be used to rephase spins independently of their velocity. In this case, the negative lobe of G_y has the same strength but twice the duration of each of the positive lobes.

can also be used to generate contrast, as in 'magnitude-contrast' angiography, which involves subtraction of two images, in one of which the signal loss associated with phase dispersion is maximized, while in the other it is minimized. A widely used approach to minimizing the effects of phase shifts is to rephase the spins independently of their velocity. This is achieved by incorporating additional gradient pulses (Fig. 8.12), a procedure that has been termed 'gradient motion rephasing' or 'gradient moment nulling'. Essentially, this procedure ensures that there is cancellation of the second term in eqn (8.4) as well as the first term. Gradient motion rephasing forms an important component of time-of-flight angiography sequences, and in conjunction with short echo times results in blood flow images of high quality. However, it is still difficult to avoid the effects of higher order motion associated with turbulent flow; the resulting signal loss raises the possibility, for example, of overestimating stenoses.

In phase-contrast angiography, the velocity-dependent phase changes associated with bipolar gradient pulses are specifically exploited in order to display the vasculature. Briefly, one data set is obtained with velocity encoding, and another is obtained either with flow compensation to suppress the velocity effects, or with reversed encoding. Vector subtraction of the two data sets should result in a residual signal from the flowing blood only. In practice, the measurements are somewhat more complicated, as the velocity encoding needs to be performed separately for each of the three (x, y, and z) axes. Therefore a total of four data sets are acquired, interleaved with each other in order to reduce adverse effects of patient motion. Phase dispersion within individual voxels needs to be minimized in order to avoid signal loss; it is therefore desirable to use small voxels.

In summary, magnetic resonance angiography is generally carried out using either time-of-flight or phase-sensitive techniques. Both approaches can be implemented in two or three dimensions. Time-of-flight methods utilize short *TE*, and short *TR* gradient-echo sequences, with a pulse angle that is selected to optimize the differential saturation between protons in the blood and in the surrounding tissue. Time-of-flight sequences use gradient motion rephasing to reduce the adverse effects of velocity-dependent phase changes. On the other hand, phase-sensitive techniques rely specifically on velocity-dependent phase changes, which are generated by means of bipolar gradient pulses. The magnitude of the phase changes can then be used to assess flow velocities.

8.7 FREQUENCY-SELECTIVE PULSES

As discussed in earlier chapters, Fourier-transform NMR exploits the use of brief, intense radiofrequency pulses to produce simultaneous excitation of signals covering a wide range of frequencies. In many circumstances, however, it is also important to excite selectively certain specific signals within a relatively narrow frequency range. For these purposes, frequency-selective radiofrequency pulses can be used.

In Section 5.6, we saw that one approach to producing frequency selectivity is to use long, weak pulses; the longer the pulse, the narrower the frequency range that it excites. Radiofrequency pulses with a rectangular intensity profile provide the simplest starting point for such an approach, but they have limitations, because their frequency response does not have sharp edges; the response has a shape that is somewhat similar to the sinc, or $(\sin x)/x$, function. As a result, the frequency selectivity is poor. Fourier analysis suggests that a much better approach would be to use pulses with a sinc profile, as these might be expected to produce a rectangular frequency response. In practice, while such pulses are now commonly used in both imaging and spectroscopy, further analysis has led to additional improvements in selectivity.

Frequency-selective pulses have widespread application, including

(1) slice selection in MRI;

(2) volume selection in localized spectroscopy;

(3) fat suppression in MRI;

(4) solvent suppression in ^{1}H MRS;

(5) magnetization transfer studies in both MRI and MRS; and

(6) selective spin-decoupling in MRS.

The incorporation of these pulses into specific sequences is illustrated in several of the following sections. Rectangular profiles in the diagrams depict 'hard' non-selective pulses, while sinc-like profiles depict 'soft' frequency-selective pulses.

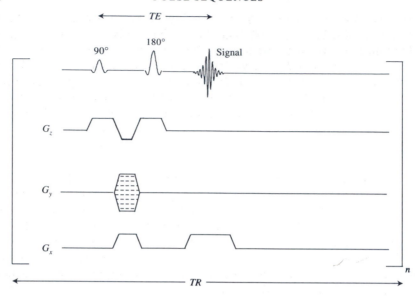

Fig. 8.13 A two-dimensional Fourier-transform imaging sequence.

8.8 THE PRODUCTION OF IMAGES

8.8.1 The two-dimensional Fourier-transform method

The generation of two-dimensional images by means of the combined use of frequency-selective pulses for slice selection, followed by spatial encoding with phase encoding and read gradients, was described in Section 5.7.2. Here, we expand on that discussion by describing some of the technical features of a two-dimensional Fourier-transform (2D FT) imaging sequence.

The sequence is shown in Fig. 8.13. One general point to note is that the gradient pulses are no longer displayed as rectangular pulses; they now have sloping sides, to indicate the finite rise and fall times that are inevitably associated with generating gradients of this type.

Slice selection is achieved by using a frequency-selective 90° pulse, applied in the presence of a field gradient G_z. This produces a variation in phase across the thickness of the slice, the effects of which are to cause a large degree of signal cancellation. Fortunately, this can be overcome by applying a reverse gradient after the pulse, as shown by the negative G_z lobe in Fig. 8.13. The strength and duration of this gradient are chosen in order to ensure that when the gradient is switched off the magnetization has returned to being in phase across the thickness of the slice. Clearly, it is necessary to wait until the end of this reversed gradient before acquiring data.

The y-dimension of the two-dimensional image is encoded by applying phase-encoding gradients G_y. As discussed in Section 5.7.2, these gradients, which are applied before, rather than during,the acquisition of data, generate a phase shift in the acquired signals, the magnitude of which reflects the location of the spins in the y-direction. It may be shown that the application of just one y-gradient does not provide enough information to generate a two-dimensional image. What is needed is to apply this gradient a number of times (say 128), gradually stepping through the gradient strength from G_{ymax} to 0 to $-G_{ymax}$. In this way, a series of data sets (in this case 128) are recorded. These gradients can be applied at the same time as the rephasing lobe of the gradient G_z.

The x-dimension of the image is encoded by acquiring data in the presence of the read gradient G_x. However, it is not possible to use the simple scheme illustrated in Fig. 8.2, for a delay is required in order to accommodate the rephasing lobe of the G_z and the phase-encoding gradient, and to allow for gradient switching effects. This delay in signal acquisition generates problems with signal loss and dephasing (see Section 7.6.2). The way to overcome these problems is to collect the signal in the form of an echo, which can recapture all of the signal except for that which is lost through T_2 or T_2^* effects. The sequence shown in Fig. 8.13 incorporates the $90°-TE/2-180°-TE/2$ spin-echo module. Another advantage of collecting data in the form of an echo is that the acquisition of both halves of the echo gives an improvement in signal-to-noise of $\sqrt{2}$, and moreover, as discussed in Section 7.6.2, there is no problem with phase correction, for the magnitude mode gives a pure absorption signal. The refocusing 180° pulse can in principle be a 'hard' non-selective pulse, but in Fig. 8.13 is shown as being slice selective.

One potential problem with acquiring data in the form of a spin-echo is that by the time the peak of the echo has been reached, the read gradient has been on for some time and has therefore resulted in considerable signal dephasing. The adverse consequences of this can be avoided by collecting the signal in the form of a gradient echo as well as a spin-echo, ensuring that the peak of the gradient echo coincides in time with the peak of the spin-echo. This is achieved by applying a pulsed field gradient G_x prior to the read gradient. This appears in Fig. 8.13 as a positive lobe applied prior to the 180° pulse; the same effect may be achieved instead by applying a negative lobe after the 180° pulse.

Two-dimensional Fourier transformation of the complete data set gives rise to a two-dimensional image. Since the sequence shown in Fig. 8.13 incorporates a spin-echo, it necessarily generates some T_2-weighting in the resulting image. However, if TE is short (e.g. less than 20 ms), the T_2-weighting will be relatively modest. T_1-Weighting could be generated simply by repeating acquisitions at intervals short enough to produce partial saturation of the signals. Alternatively, an inversion recovery ($180°-TI-90°$) radio-frequency pulse module could be incorporated into the sequence. More generally, it is possible to superimpose a variety of different types of

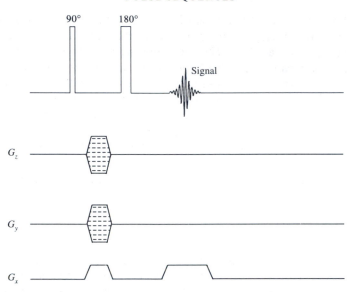

Fig. 8.14 A three-dimensional Fourier-transform spin-echo imaging sequence. As discussed in Section 8.8.4, however, (see also Fig. 8.15) gradient echoes are normally preferred for three-dimensional acquisition.

weighting simply by incorporating appropriate additional radiofrequency and/or gradient pulses into the sequence.

The 2D FT method illustrated in Fig. 8.13 is probably the most widely used of the magnetic resonance imaging methods. However, a range of other approaches are feasible; indeed the first proposal (Lauterbur 1973) used an alternative method known as projection–reconstruction, which involves simultaneous application of the G_x- and G_y-gradients. This gradually fell out of favour, however. In part this was because of the effects of gradient imperfections, which in projection–reconstruction tend to give blurring of images, whereas in 2D FT imaging they produce distortions, which are more readily tolerated. However, the projection–reconstruction method enables shorter echo delays to be used, and is receiving renewed interest.

8.8.2 Three-dimensional imaging

The extension from two-dimensional to three-dimensional imaging is fairly straightforward, at least in principle. First, the slice-selective pulse is replaced by a non-selective pulse to excite a three-dimensional region, as shown in Fig. 8.14. (Alternatively, the pulse may remain selective, but for the excitation of a thick slab rather than for excitation of the whole object or a thin slice within the object.) Secondly, phase-encoding gradients are applied in both the y- and z-directions. The problem is that this will require a large array of data; for example it may be necessary to acquire not 128 scans, as in two-dimensional studies, but 128×128 scans, depending upon the required resolution in the various directions. This imposes two difficulties; the first is the total data acquisition time that is required in order to collect this large number of scans, and the second is the large amount of data for Fourier transformation and subsequent analysis. Fortunately, the combination of rapid imaging techniques (see Section 8.8.4) and ever-increasing computer power is enabling these problems to be overcome, and three-dimensional imaging is becoming increasingly widespread. There still remains a fairly formidable problem of the time required to interpret the huge bank of information that such data sets provide!

8.8.3 Multi-slice imaging

Before the introduction of true three-dimensional imaging, multi-slice imaging provided a means of scanning through three dimensions reasonably rapidly and efficiently. We have seen that TR is generally much greater than the time needed to acquire data; TR may be of the order of 1 s, while the sequence required for data acquisition may last 100 ms. Instead of allowing the NMR instrument to be inactive for the residual 900 ms, it is possible to use this period for the successive collection of data from different selected slices. The different slices are excited in turn by appropriate selection of the frequencies of the frequency-selective pulses. If eight slices are to be observed, it is common practice to irradiate the slices in an order such as [1,4,7,2,5,8,3,6] rather than [1,2,3,4,5,6,7,8]. This avoids problems resulting from the fact that imperfections in slice-selection can generate unwanted magnetization in adjacent slices. The interleaving of data acquisition through multi-slice imaging provides a simple means of acquiring data in all three dimensions, and is widely used in clinical imaging.

8.8.4 Rapid imaging techniques

There is extensive interest in rapid imaging techniques, and this will undoubtedly increase in the coming years. Here, we discuss the two main approaches that have been adopted. The first is a relatively simple extension of conventional imaging; essentially it involves shortening the TE and TR times as much as possible. In order to do this, it is necessary to eliminate the 180° refocusing pulse, and instead rely on the use of gradient echoes, as illustrated in Fig. 8.15. With conventional technology, gradient-echo times as low as 2–3 ms have been achieved, with TR values of 3–5 ms. The total acquisition time is, of course, equal to TR multiplied by the number of phase-encoding steps. When using short TR values, there is obviously a danger of saturating the signals, and therefore the pulse angle has to be reduced according to the

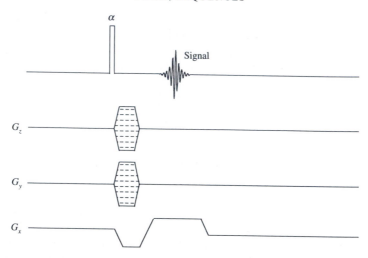

Fig. 8.15 A three-dimensional gradient-echo imaging sequence.

equations and diagrams referred to in Section 7.3. Rapid imaging techniques have acquired a number of acronyms, but the most widespread (and perhaps the most appropriate) is FLASH, which stands for fast low-angle shot imaging (Haase *et al.* 1976). There are a number of variants of this approach to rapid imaging, with a range of differing contrast characteristics (Frahm *et al.* 1992). One of the features of such techniques is that they provide a means of acquiring full three-dimensional data sets in a reasonable amount of time, and with ever-increasing computer power it seems likely that this type of data acquisition will become increasingly common. An important advantage of acquiring a full three-dimensional data set is that the data can be reformatted in order to display planes in any orientation that may be required.

The second rapid imaging approach is termed echo-planar imaging (Mansfield and Pykett 1978; Stehling *et al.* 1991). This method provides a 'snapshot' facility, in the sense that a complete image can be obtained following one, initial excitation pulse, in a time as little as 40 ms. Echo-planar imaging imposes considerable demands upon machine performance, and until recently has not been widely available. However, for a number of reasons, its availability is increasing considerably. This is partly because of various technical developments and (perhaps more importantly) also because of the increasing awareness of the value of a 'snapshot' imaging facility, not just for freezing the effects of body motion, but also for investigation of time-dependent effects such as the processes associated with activation of brain function.

Figure 8.16 shows one version of the echo-planar imaging sequence. It may be seen that it requires rapid switching of both the read and phase-encoding gradients in order to encode for two dimensions during the course of a single free induction decay. It is not straightforward to visualize how the spins are

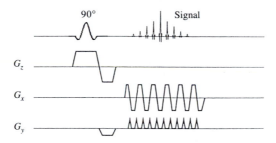

Fig. 8.16 An echo-planar imaging sequence.

influenced by this array of switched gradients. One approach is to utilize the concept of k-space, which was introduced in Section 5.7. Figure 8.17(a) shows the trajectory through k-space for conventional two-dimensional Fourier-transform imaging. Each horizontal line represents the data acquired in the presence of a read gradient. The full two-dimensional data set is obtained by incrementing the phase-encoding gradient in equal steps, and this is represented in the diagram by stepping from one horizontal line to the next. The full traversal of k-space provides the information which, on Fourier transformation, generates a full two-dimensional image. The trajectory through k-space differs in echo-planar imaging, as illustrated in Fig. 8.17(b), the complete two-dimensional trajectory being covered in a single acquisition.

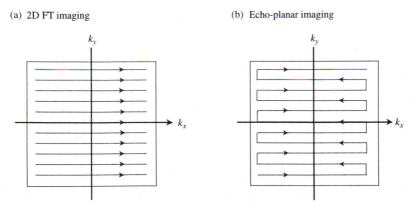

Fig. 8.17 Trajectories through k-space for (a) two-dimensional Fourier-transform imaging, and (b) for echo-planar imaging. (Adapted from Stehling *et al.* 1991.)

8.9 LOCALIZED SPECTROSCOPY

Numerous techniques have been proposed for obtaining spectra from well-defined localized regions of the body. Of these, only a handful remain in widespread use, and here we discuss the most commonly used techniques.

8.9.1 The use of surface coils

The initial method of spectral localization involved the spatial selectivity of surface coils (Ackerman *et al.* 1980). Their selectivity is based on the B_1 field distribution of this type of radiofrequency coil. While the localization provided by such coils has proved to be useful for superficial tissues, in particular for skeletal muscle, it should be appreciated that the sensitive volume is not very well defined, and interpretation of some of the data obtained with surface coil localization alone therefore needs to be treated with some caution. Phosphorus-31 spectroscopy studies of skeletal muscle tend to be reliable, because there is relatively little 'contaminating' signal from other tissues. However, if surface coil localization is used for examination of tumours, there is always the possibility that skeletal muscle, with its particularly high phosphocreatine signal, may contribute to the spectra, and so spectral changes associated, for example, with therapy may conceivably reflect changing contributions from such neighbouring muscle tissue. Of course, surface coils are advantageous for examination of superficial tissue not just because of their localizing properties, but also because they provide higher sensitivity for such studies than enveloping coils. Therefore, there is often a role for combining surface coil detection with additional localization techniques of the type described in the following sections. Such studies commonly use an enveloping coil for transmission and the surface coil for signal reception (see Section 7.4.3).

The spatial selectivity of surface coils relies on the characteristics of their B_1 field distribution. It is possible to build on this concept in order to provide a basis for imaging. Consider, for example, a transmitter coil that generates a B_1 field varying linearly along a given direction; the pulse angle along this direction would then also vary linearly. By examining the variation of signal intensity with increasing pulse width or B_1 field strength, the location of the nuclei giving rise to the spins could then be determined; the linear dependence of the B_1 field is analogous in some respects to the linear field gradients G_x, G_y and G_z that are used in conventional one-dimensional imaging. This use of B_1 field gradients provides the basis for a technique termed 'rotating frame imaging' (Hoult 1979). However, the technique has not found widespread application, primarily because gradients in the radiofrequency B_1 field are much less easily generated and controlled than the static field gradients that are used more conventionally.

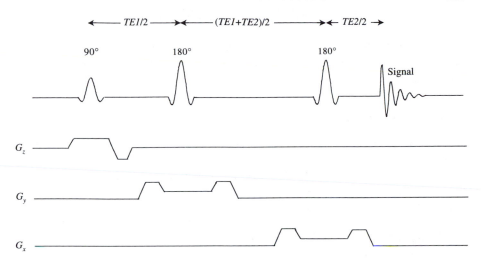

Fig. 8.18 Pulse sequence for PRESS (point-resolved spectroscopy).

8.9.2 Single voxel techniques

A variety of techniques based on static field gradients have been developed for obtaining spectra from single volume elements (or voxels) of interest. Here, we discuss the three techniques that are most commonly used.

8.9.2.1 *PRESS*

PRESS, an acronym for point-resolved spectroscopy, uses the following pulse sequence (Ordidge *et al.* 1985; Bottomley 1987):

$$90°-TE1-180°-(TE1 + TE2)-180°-TE2-\text{Acquire}$$

as illustrated in Fig. 8.18. Each of the three radiofrequency pulses is slice selective, selection being along the x-axis for one pulse, along the y-axis for the second pulse, and along the z-axis for the third. The first pair of pulses generates an echo at time $TE1$ after the first 180° pulse, but only from those spins that are located in the region intersected by the two selected slices. A second echo is generated at time $TE2$ after the second 180° pulse, but only from those spins that have been excited by all three of the radiofrequency pulses. This corresponds to a single volume of interest defined by the inter-section of the selected slices. Spoiler gradients are incorporated into the sequence in order to avoid the possibility of observing unwanted signals from spins that experience only one or two of the radiofrequency pulses. A potential problem with this sequence is that the echo time TE is necessarily quite large in order to accommodate the radiofrequency pulses and spoiler gradients, and the method is therefore not very well suited to the detection of signals with short T_2 values or complex coupling patterns for which there is severe signal

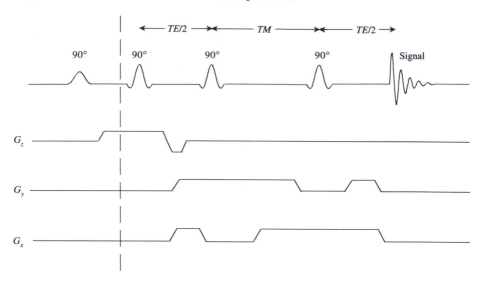

Fig. 8.19 Pulse sequence for STEAM (stimulated echo acquisition mode). A solvent suppression 'module' is shown to the left of the dashed line.

loss at long echo times. In practice, therefore, this technique is little used for ^{31}P MRS, but quite extensively used for detecting the principal singlet and doublet resonances seen in ^{1}H MRS.

For ^{1}H spectroscopy studies, the PRESS sequence is normally preceded by a solvent suppression 'module'. A simple means of selectively suppressing the water peak is shown in the first part of the sequence shown in Fig. 8.19. A frequency-selective 90° pulse is applied such that significant excitation only occurs within a small chemical shift range that includes the water peak. A spoiler gradient is then applied in order to destroy the net magnetization in the xy-plane. As a result of this, the water protons have no net magnetization along any orientation, whereas the magnetization of protons with different chemical shifts remains along the z-axis.

8.9.2.2 *STEAM*

STEAM is an acronym for the stimulated echo acquisition mode sequence (Frahm *et al.* 1987). The sequence (preceded by solvent suppression) is shown in Fig. 8.19, and can be written as:

$$90°\text{-}TE/2\text{-}90°\text{-}TM\text{-}90°\text{-}TE/2\text{-}\text{Acquire}$$

where *TM* is a delay during which the magnetization is longitudinal rather than in the transverse plane. In a similar manner to the PRESS sequence, the three slice-selective radiofrequency pulses are applied in the presence of x-, y-, and z-gradients in order to generate signal from a single volume of interest. The first 90° pulse generates transverse magnetization which is converted by the

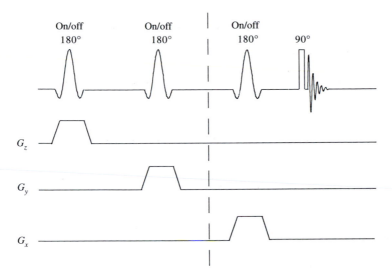

Fig. 8.20 Pulse sequence for ISIS (image selected *in vivo* spectroscopy).

second pulse back into longitudinal magnetization. The effect of the third pulse is to regenerate transverse magnetization, which reappears as a so-called stimulated echo. As with the PRESS sequence, spoiler gradients are incorporated into the sequence in order to eliminate unwanted signals. It may be shown that the stimulated echo only recaptures half of the available signal, and therefore in principle STEAM only has half the sensitivity of PRESS. However, it can generally achieve shorter echo times, primarily because the interval between the second and third pulses does not contribute to the effective echo time. Thus STEAM is useful for detecting metabolites with short T_2 values or with complex coupling patterns such as glutamate and glutamine in ^{1}H MRS. It is of interest to note that the phase modulation characteristics of spin-coupled signals differ when using the STEAM sequence from those obtained using conventional spin echoes (Ernst and Hennig 1991).

8.9.2.3 *ISIS*

In contrast to the above two methods, ISIS (image selected *in vivo* spectroscopy; Ordidge *et al.* 1986) does not involve echo formation. Instead, it uses slice-selective 180° pulses to prepare the magnetization prior to exciting the entire magnetization of the object with a non-selective 90° pulse. To illustrate the method, consider first a one-dimensional ISIS experiment. This is illustrated by the sequence shown to the right of the dashed line in Fig. 8.20. The first scan involves the collection of a free induction decay after a single 90° pulse (i.e. with the 180° pulse off), while for the second scan an additional slice-selective 180° pulse is applied. This causes inversion of the signal from within the selected slice, but no change in the signal detected from outside this slice. Subtraction of one free induction decay from the other therefore produces signal from within the selected slice, signals from outside this slice cancelling each other out. Extension to three dimensions, illustrated by the complete sequence shown in Fig. 8.20, requires the use of a cycle of eight on/off combinations rather than two, utilizing three 180° pulses with slice selection along each of the x-, y-, and z-axes. Localization is achieved by appropriate addition and subtraction of the data acquired from the cycle of eight free induction decays.

Since ISIS involves data acquisition immediately after the 90° pulse, it is well suited for the detection of signals that decay rapidly (i.e. signals with short T_2 values or complex coupling patterns), and is commonly used for ^{31}P MRS. However, it has the disadvantage that motion during the accumulation of data can result in severe localization errors since signals from outside the region of interest will not subtract well. Furthermore, while localized 'shimming' on the region of interest can be carried out with the PRESS and STEAM sequences, this is not possible with ISIS, since each individual FID includes signal from outside the region of interest. The development of modifications such as additional outer volume suppression (Connelly *et al.* 1988) have helped to overcome these problems, but ISIS has nevertheless found relatively little application in ^{1}H MRS.

8.10 SPECTROSCOPIC IMAGING

The above localization techniques all suffer from the disadvantage that only one region (or, with some modifications, a few regions) can be monitored at any given time. The observation of multiple regions with these techniques therefore requires sequential acquisition of data from the different regions of interest. When it is desirable to monitor the metabolic state of numerous regions, it would clearly be preferable to detect all regions simultaneously, just as in conventional imaging. Chemical shift (or spectroscopic) imaging techniques[1] provide a means of doing this, (Brown *et al.* 1982; Maudsley *et al.* 1983), the main problem being that, just as for single voxel techniques, the spatial resolution is necessarily fairly crude because of constraints imposed by signal-to-noise; since the metabolites are present at only millimolar concentrations (in contrast to water, which is present in tissues at about 40 M), adequate signal-to-noise ratios are obtained only from relatively large volumes, with volume resolution of at best about 1 cc for clinical studies.

A simple spectroscopic imaging sequence is shown in Fig. 8.21. In order to maintain the chemical shift information, it is necessary to collect the free induction decay in the absence of any field gradients. Therefore all the spatial encoding needs to be carried out prior to acquisition of the decay.

---

[1] We use the term spectroscopic imaging here, as chemical shift imaging is now often used in a more specific sense to refer to MRI methods that provide separate images of water and fat.

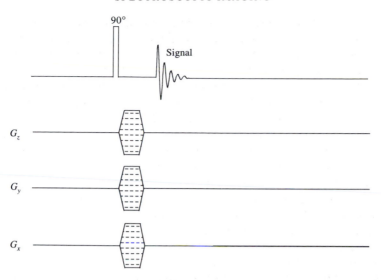

Fig. 8.21 Spectroscopic imaging sequence.

This is achieved by collecting a series of free induction decays following the application of phase-encoding gradients, which can be applied along one, two, or three directions depending upon whether a one-, two-, or three-dimensional metabolic map is required. Fourier transformation of the complete data set generates an array of spectra which can be displayed in the form of 'metabolic images' showing the spatial distribution of signals from each of the metabolites. Spectroscopic imaging is an efficient technique in that signal is acquired from many regions simultaneously; however, in order to obtain this information, it is necessary to accumulate data for a long time, because of the large number of phase-encoding steps that are required, particularly in the case of three-dimensional metabolic mapping.

Single volume techniques and spectroscopic imaging methods each have their advantages and disadvantages. Using single volume techniques, shimming can be carried out over the specific volume of interest, which may produce superior results to the more global shimming associated with chemical shift imaging procedures. In addition, shorter measuring times can generally be achieved with single volume techniques. While the volumes observed with the PRESS, STEAM, and ISIS techniques are fairly well defined, it is important, however, to appreciate that these are influenced by the precise slice profile characteristics of the 90° and 180° pulses, and also that compounds with differing chemical shifts will be localized to slightly different regions. This is because the slice-selective pulses achieve their spatial selectivity on the basis of the frequencies of the signals, and do not distinguish between frequency differences associated with chemical shifts and those associated with differing spatial locations. Typically, this effect will be small for ^{1}H MRS, but may

produce a relative displacement as large as 1 cm out of 5 cm for localized
^{31}P spectroscopy. The phase-encoding techniques used for spectroscopic
imaging do not suffer from this artefact. However, one problem with spectroscopic imaging is that the signal attributed to each volume element contains
significant contributions from neighbouring volume elements, and this has
to be taken into consideration when looking for highly localized spectral
abnormalities.

In conclusion, single volume techniques and spectroscopic imaging each
have their characteristic advantages and disadvantages, and it seems very likely
that both will continue to play extensive roles in the spectroscopy of living
systems.

8.11 SPIN–SPIN COUPLING AND SPECTRAL EDITING

Numerous techniques have been developed for improving the quality and
aiding the interpretation of NMR spectra. Many of these methods are based
on the phenomenon of spin–spin coupling and on the behaviour of coupled
lines in spin-echo sequences. In this section, we briefly discuss some of these
methods.

The phenomenon of spin–spin coupling was discussed in some detail
in Section 6.2. To extend that discussion, consider firstly a ^{13}C study of
acetaldehyde (CH_3CHO) in which the CH carbon is ^{13}C-labelled. This carbon
is spin-coupled to the CH proton, with a coupling constant of about 100 Hz.
If a 90° pulse is applied at the ^{13}C frequency, then immediately after the
pulse the ^{13}C nuclear spins will be aligned in the $x'y'$-plane, as shown in
Fig. 8.22(a). If we ignore any dephasing due to relaxation or field inhomogeneities, the nuclear spins will then separate into two groups (because of the
spin–spin coupling interaction), precessing at different frequencies ('fast' and
'slow') as shown in Fig. 8.22(b). The two frequencies differ by the value of
the coupling constant, which in this case is 100 Hz. If a 180° pulse is applied
at a time $TE/2$, then the coupled spins will behave as previously described
in Section 8.3; after the pulse they will continue to precess at the same frequency and form an echo after a further time $TE/2$ (Figs 8.22(c) and (d)). They
do not, therefore, undergo the process known as phase modulation. Suppose,
however, that a 180° pulse is also applied at time $TE/2$ to the CH protons,
causing the protons effectively to interchange their spin states. Because the
^{13}C carbons are coupled to these protons, the precessional frequencies also
interchange, i.e. the 'fast' carbons become 'slow', and vice versa (Fig. 8.22(e)).
As a result, after a further time $TE/2$, the carbons do not necessarily return
to being in phase with each other (Fig. 8.22(f)); they may, for example, be
90° or 180° out of phase with each other, depending on the value of TE in
relation to the coupling constant.

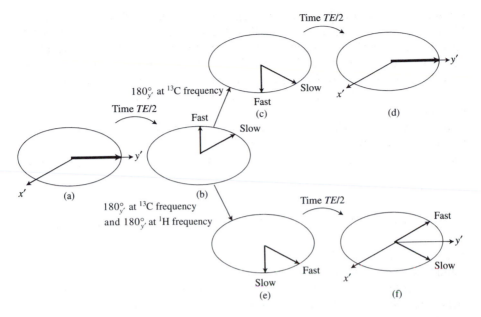

Fig. 8.22 Diagram illustrating behaviour of spin–spin coupled signals in a spin-echo experiment.

This behaviour is termed 'phase modulation'; even for a simple doublet such as this, the response on Fourier transformation is not totally straightforward, for the two components may have different phases, not only with respect to the rest of the spectrum, but also with respect to each other. For more complex spin systems, very complicated patterns can emerge, and there is a tendency for destructive interference to take place between the various components of the multiplets, causing a significant loss of signal intensity.

Now let us move from ^{13}C MRS to the ^{1}H MRS characteristics of unlabelled acetaldehyde, in particular the CH$_3$ protons, which are coupled to the CH proton. If a conventional spin-echo sequence is performed, then the situation for the CH protons will be analogous to that shown in Fig. 8.22(e) and (f); the 180° pulse irradiates the whole ^{1}H spectrum, and therefore inverts not only the CH$_3$ protons but also the CH proton to which they are coupled. Therefore the CH$_3$ protons undergo the same type of phase-modulation process, the difference being that because the coupling constant is very much smaller (about 7 Hz), longer *TE* values are needed in order to observe the equivalent amount of modulation. In fact, it is easy to show that after a time $1/2J$, i.e. about 140 ms, the CH$_3$ doublet is inverted relative to uncoupled spins. The behaviour of the CH$_3$ protons of lactate is just the same as for

acetaldehyde, and this is why a *TE* value of 135 ms is commonly used in
^{1}H MRS studies of tissue metabolism; at this *TE* value the CH signals of
lactate are inverted, as was shown in Fig. 3.4(a). In the case of lactate or
acetaldehyde, the phase modulation may be abolished by applying selective-
decoupling irradiation to the CH protons during the time *TE*. Alternatively,
an additional selective 180° pulse (straddling the broadband 180° pulse) can
be applied to the CH protons. Another approach to abolishing the phase
modulation is to use radiofrequency pulses whose frequency selectivity is such
that they do not irradiate the CH protons at all.

From the above discussion, it should be apparent that 'spectral editing',
i.e. selective detection of specified signals, should be achievable by appropriate
manipulation of the phase modulation of coupled spins. For example, if we
consider the lactate CH_3 doublet, which is centred at 1.32 ppm, a potential
problem with unambiguous detection of this signal is that it overlaps with
the lipid signal, which can be quite intense. If we carry out two different
spin-echo (*TE* = 135 ms) sequences, with phase modulation present in one but
abolished in the other, then subtraction of the two sets of data should
eliminate the lipid signal (because it should be the same in the two cases),
leaving the lactate signal (which is negative in one of the sequences and positive
in the other).

The same type of idea can be put to particularly good use in ^{13}C-labelling
studies. A proton that is adjacent to a ^{13}C-labelled carbon will be split into
two components as a result of spin–spin coupling with the ^{13}C nucleus, and
will undergo phase modulation in ^{1}H spectroscopy. If a 180° pulse is applied
to the ^{13}C nuclei at time *TE*/2, then this phase modulation will be abolished.
Therefore subtraction of data obtained with and without the ^{13}C 180° pulse
should reveal those protons that are adjacent to ^{13}C nuclei. Effectively this
means that the ^{13}C-labels are being detected through ^{1}H MRS, the important
feature of this being that ^{1}H MRS is much more sensitive than ^{13}C MRS. The
effectiveness of this approach to metabolic studies *in vivo* was demonstrated
by Rothman *et al.* (1985) in their studies of brain metabolism, and since then
numerous MRS studies of tissue metabolism have taken advantage of the
additional information that spectral editing techniques can provide.

8.11.1 Spin decoupling

In Section 6.2, it was pointed out that multiplets could be collapsed into
single lines by applying irradiation at the resonance frequency of the nuclei
to which the spins of interest are coupled. This process of spin decoupling
involves the application of irradiation continuously during the acquisition of
the free induction decay, as illustrated in Fig. 8.23(a). Such use of a second
irradiation channel is commonly termed double irradiation, and denoted by
a radiofrequency field B_2. Spin decoupling is most commonly used in ^{13}C
NMR. The decoupling irradiation can be applied at a selected proton frequency;

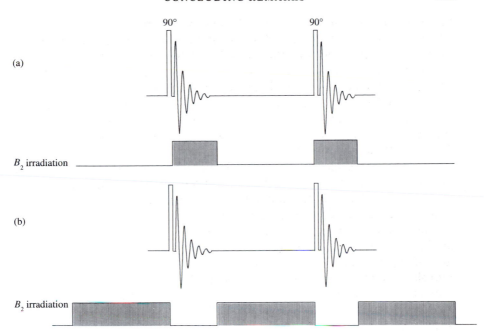

Fig. 8.23 Double irradiation: (a) applied during each data acquisition, and (b) prior to each acquisition.

alternatively 'broadband' decoupling can be used, whereby the whole of the ^{1}H spectral range is irradiated. This can involve considerable irradiating power, and a number of innovative strategies have been described that enable broad-band decoupling to be carried out without excessive power requirements (see Freeman 1988). In ^{13}C NMR studies, it is also common practice to saturate the ^{1}H signals prior to acquisition of the ^{13}C signals, as shown in Fig. 8.23(b); this can lead to large enhancements in the intensities of the ^{13}C signals through the nuclear Overhauser effect (see Section 6.4). Therefore, ^{13}C studies commonly utilize continuous ^{1}H irradiation, both between and during the acquisition of consecutive free induction decays. For similar reasons, ^{1}H irradiation is increasingly used in ^{31}P MRS and, as mentioned above, selective irradiation of specific protons can also be used for spectral editing in ^{1}H spectroscopy.

8.12 CONCLUDING REMARKS

At the time of writing, we are approaching the 50th anniversary of the initial successful NMR measurements of Bloch, Hansen and Packard; and of Purcell,

Torrey, and Pound. Throughout this 50 years, there have been continual remarkable developments in NMR, some of which have surprised even the greatest of NMR scientists. Erwin Hahn, for example, has stated (Hahn 1990):

... I never believed MRI would work, like Rutherford, who said that anyone who believed nuclear radioactivity would be useful 'is talking moonshine'. However, I was only one of many unbelievers. Another infidel in particular was Anatole Abragam, a distinguished French physics researcher in magnetic resonance. He notes in his autobiography (1989) that French clinicians began to buy his book, *Principles of magnetic resonance*, thinking it would enlighten them in their speciality of MRI. The French Society of Radiology wanted to award Abragam a medal in spite of the fact that he told them he hadn't contributed to MRI and didn't believe it would work.

The development of new pulse sequences, and the discovery of new applications for these sequences, has played a central role in the inexorable progress of NMR. It can only be assumed that this progress will continue, and that there are many more surprises around the corner.

REFERENCES

Ackerman, J. J. H., Grove, T. H., Wong, G. G., Gadian, D. G., and Radda, G. K. (1980). Mapping of metabolites in whole animals by ^{31}P NMR using surface coils. *Nature*, **283**, 167–70.

Bottomley, P. A. (1987). Spatial localization in NMR-spectroscopy in vivo. *Ann. N.Y. Acad. Sci.*, **508**, 333–48.

Bradley, W. G. jun. (1992). Flow phenomena. In *Magnetic resonance imaging* (eds D. D. Stark and W. G. Bradley jun.), pp. 253–98. Mosby Year Book, St. Louis.

Brown, T. R., Kincaid, B. M., and Ugurbil, K. (1982). NMR chemical shift imaging in three dimensions. *Proc. Natl. Acad. Sci. USA*, **79**, 3523–6.

Carr, H. Y. and Purcell, E. M. (1954). Effects of diffusion on free precession in nuclear magnetic resonance experiments. *Phys. Rev.*, **94**, 630–8.

Connelly, A., Counsell, C., Lohman, J. A. B. and Ordidge, R. J. (1988). Outer volume suppressed image related in vivo spectroscopy (OSIRIS), a high-sensitivity localization technique. *J. Magn. Reson.*, **78**, 519–25.

Ernst, T. and Hennig, J. (1991). Coupling effects in volume selective ^{1}H spectroscopy of major brain metabolites. *Magn. Reson. Med.*, **21**, 82–96.

Frahm, J., Merboldt, K. D., and Hanicke, W. (1987). Localized proton spectroscopy using stimulated echoes. *J. Magn. Reson.*, **72**, 502–8.

Frahm, J., Gyngell, M. L., and Hanicke, W. (1992). Rapid scan techniques. In *Magnetic resonance imaging* (eds D. D. Stark and W. G. Bradley jun.), Mosby Year Book, St. Louis.

Freeman, R. (1988). *A handbook of magnetic resonance*. Longman Scientific and Technical, Harlow.

Haase, A., Frahm, J., Matthaei, D., Hanicke, W., and Merboldt, K-D. (1986). FLASH imaging. Rapid NMR imaging using low flip-angle pulses. *J. Magn. Reson.*, **67**, 258–66.

Hahn, E. L. (1950). Spin echoes. *Phys. Rev.*, **80**, 580–94.

Hahn, E. L. (1990). MRI in retrospect. *Phil. Trans. R. Soc. Lond. A*, **333**, 403–11.

Hoult, D. I. (1979). Rotating frame zeugmatography. *J. Magn. Reson.*, **33**, 183–97.

Lauterbur, P. C. (1973). Image formation by induced local interactions: examples employing nuclear magnetic resonance. *Nature*, **242**, 190–1.

Le Bihan, D. and Turner, R. (1992). Diffusion and perfusion. In *Magnetic resonance imaging* (eds. D. D. Stark and W. G. Bradley jun.), pp. 335–71. Mosby Year Book, St. Louis.

Mansfield, P. and Pykett, I. L. (1978). Biological and medical imaging by NMR. *J. Magn. Reson.*, **29**, 355–73.

Masaryk, T. J., Lewin, J. S., and Laub, G. (1992). Magnetic resonance angiography. In *Magnetic resonance imaging* (eds. D. D. Stark and W. G. Bradley jun.), pp. 299–334. Mosby Year Book, St. Louis.

Maudsley, A. A., Hilal, A. K., Perman, W. H., and Simon, H. E. (1983). Spatially resolved high-resolution spectroscopy by 'four-dimensional' NMR. *J. Magn. Reson.*, **51**, 147–52.

Meiboom, S. and Gill, D. (1958). Modified spin-echo method for measuring nuclear relaxation times. *Rev. Sci. Instrum.*, **29**, 688–91.

Ordidge, R. J., Bendall, M. R., Gordon, R. E., and Connelly, A. (1985). Volume selection for in-vivo biological spectroscopy. In *Magnetic resonance in biology and medicine* (eds G. Govil, C. L. Khetrapal, and A. Saran), pp. 387–97. Tata McGraw-Hill, New Delhi.

Ordidge, R. J., Connelly, A., and Lohman, J. A. B. (1986). Image-selected in vivo spectroscopy (ISIS). A new technique for spatially selective NMR spectroscopy. *J. Magn. Reson.*, **66**, 283–94.

Rothman, D. L., Behar, K. L., Hetherington, H. P., Bendall, M. R., Petroff, O. A. C., and Shulman, R. G. (1985). ^{1}H observe ^{13}C decoupled spectroscopic measurements of lactate and glutamate in the rat brain in vivo. *Proc. Natl. Acad. Sci. USA*, **82**, 1633–7.

Stehling, M. K., Turner R., and Mansfield, P. (1991). Echo-planar imaging: magnetic resonance imaging in a fraction of a second. *Science*, **254**, 43–50.

Index